T0200688

TOWARDS A PHILOSOPHY OF REAL MATHEMATICS

In this ambitious study, David Corfield attacks the widely held view that it is the nature of mathematical knowledge which has shaped the way in which mathematics is treated philosophically, and claims that contingent factors have brought us to the present thematically limited discipline. Illustrating his discussion with a wealth of examples, he sets out a variety of new ways to think philosophically about mathematics, ranging from an exploration of whether computers producing mathematical proofs or conjectures are doing real mathematics, to the use of analogy, the prospects for a Bayesian confirmation theory, the notion of a mathematical research programme, and the ways in which new concepts are justified. His highly original book challenges both philosophers and mathematicians to develop the broadest and richest philosophical resources for work in their disciplines, and points clearly to the ways in which this can be done.

DAVID CORFIELD holds a Junior Lectureship in Philosophy of Science at the University of Oxford. He is co-editor (with Jon Williamson) of *Foundations of Bayesianism* (2001), and he has published articles in journals including *Studies in History and Philosophy of Science* and *Philosophia Mathematica*.

TOWARDS A PHILOSOPHY
OF REAL MATHEMATICS

DAVID CORFIELD

CAMBRIDGE
UNIVERSITY PRESS

CAMBRIDGE UNIVERSITY PRESS
Cambridge, New York, Melbourne, Madrid, Cape Town, Singapore, São Paulo

Cambridge University Press
The Edinburgh Building, Cambridge CB2 2RU, UK

Published in the United States of America by Cambridge University Press, New York

www.cambridge.org
Information on this title: www.cambridge.org/9780521817226

First published 2003
Reprinted 2004

A catalogue record for this publication is available from the British Library

ISBN-13 978-0-521-81722-6 hardback
ISBN-10 0-521-81722-6 hardback

Transferred to digital printing 2006

From the east to western Ind,
No jewel is like Rosalind.

Contents

Preface

I should probably not have felt the desire to move into the philosophy of mathematics had it not been for my encounter with two philosophical works. The first of these was Imre Lakatos's *Proofs and Refutations* (1976), a copy of which was thrust into my hands by a good friend Darian Leader, who happens to be the godson of Lakatos. The second was an article entitled 'The Uses and Abuses of the History of Topos Theory' by Colin McLarty (1990), a philosopher then unknown to me. What these works share is the simple idea that what mathematicians think and do should be important for philosophy, and both express a certain annoyance that anyone could think otherwise.

Finding a post today as a philosopher of mathematics is no easy task. Finding a post as a philosopher of mathematics promoting change is even harder. When a discipline is in decline, conservatism usually sets in. I am, therefore, grateful beyond words to my PhD supervisor, Donald Gillies, both for his support over the last decade and for going to the enormous trouble of applying for the funding of two research projects, succeeding in both, and offering one to me. The remit of the project led me in directions I would not myself have chosen to go, especially the work reported in chapters 2 and 3, and I rather think chapters 5 and 6 as well, but this is often no bad thing. I am thus indebted to the Leverhulme Trust for their generous financial support. Thanks also to Jon Williamson, the other fortunate recipient, for discussions over tapas.

Colin McLarty has provided immense intellectual and moral support over the years, and also arranged a National Endowment of the Humanities Summer Seminar where sixteen of us were allowed the luxury of talking philosophy of mathematics for six weeks in the pleasant surroundings of Case Western Reserve University. My thanks to the NEH and to the other participants for making it such an enjoyable experience.

I should also like to acknowledge the helpful advice of Ronnie Brown, Jeremy Butterfield, James Cussens, Matthew Donald, Jeremy Gray, Colin

Howson, Mary Leng, Penelope Maddy, Stephen Muggleton, Madeline Muntersbjorn, Jamie Tappenden, Robert Thomas and Ed Wallace. This book could only have benefited from greater exposure to the intellectual ambience of the History and Philosophy of Science Department in Cambridge, where the writing was finished. Unfortunately time was not on my side. I only hope a little of the spirit of the department has trickled through into its pages.

Hilary Gaskin at Cambridge University Press has smoothed the path to publication. Four of the chapters are based on material published elsewhere. Chapter 5 is based on my chapter in Corfield and Williamson 2001, *Foundations of Bayesianism*, Kluwer. Chapters 7 and 9 are based on papers of the same title in *Studies in the History and Philosophy of Science*, 28(1): 99–121 and 32(3): 507–33. Chapter 8 is likewise based on my article in *Philosophia Mathematica* 6: 272–301. I am grateful to Kluwer, Elsevier and Robert Thomas for permission to publish them.

I should like to thank J. Scott Carter and Masahico Saito for kindly providing me with the figure displayed on the cover. It shows one of the ingenious ways they have devised of representing knotted surfaces in four-dimensional space. In chapter 10 we shall see how this type of representation permits diagrammatic calculations to be performed in higher-dimensional algebra.

Love and thanks to Oliver, Kezia and Diggory for adding three more dimensions to my life beyond the computer screen, and to my parents for all their support. This book I dedicate to Ros for fourteen years of sheer bliss.

CHAPTER I

Introduction: a role for history

To speak informatively about bakery you have got to have put your hands in the dough.
(Diderot, *Oeuvres Politiques*)

The history of mathematics, lacking the guidance of philosophy, has become *blind*, while the philosophy of mathematics, turning its back on the most intriguing phenomena in the history of mathematics, has become *empty*.
(Lakatos, *Proofs and Refutations*)

I.I REAL MATHEMATICS

To allay any concerns for my mental health which the reader may be feeling if they have come to understand from the book's title that I believe mathematics based on the real numbers deserves singling out for philosophical treatment, let me reassure them that I mean no such thing. Indeed, the glorious construction of complex analysis in the nineteenth century is a paradigmatic example of what 'real mathematics' refers to.

The quickest way to approach what I *do* intend by such a title is to explain how I happened upon it. Several years ago I had been invited to talk to a philosophy of physics group in Cambridge and was looking for a striking title for my paper where I was arguing that philosophers of mathematics should pay much closer attention to the way mathematicians do their research. Earlier, as an impecunious doctoral student, I had been employed by a tutorial college to teach eighteen-year-olds the art of jumping through the hoops of the mathematics 'A' level examination. After the latest changes to the course ordained by our examining board, which included the removal of all traces of the complex numbers, my colleagues and I were bemoaning the reduction in the breadth and depth of worthwhile content on the syllabus. We started playing with the idea that we needed a campaign for the teaching of real mathematics. For the non-British and those with no interest in beer, the allusion here is to the Campaign for Real Ale (CAMRA), a movement dedicated to maintaining traditional brewing

I

techniques in the face of inundation by tasteless, fizzy beers marketed by powerful industrial-scale breweries. From there it was but a small step to the idea that what I wanted was a Campaign for the Philosophy of Real Mathematics. Having proposed this as a title for my talk, it was sensibly suggested to me that I should moderate its provocative tone, and hence the present version.

It is generally an indication of a delusional state to believe without first checking that you are the first to use an expression. The case of 'real mathematics' would have proved no exception. In the nineteenth century Kronecker spoke of 'die wirkliche Mathematik' to distinguish his algorithmic style of mathematics from Dedekind's postulation of infinite collections. But we may also find instances which stand in need of no translation. Listen to G. H. Hardy in *A Mathematician's Apology*:

It is undeniable that a good deal of elementary mathematics – and I use the word 'elementary' in the sense in which professional mathematicians use it, in which it includes, for example, a fair working knowledge of the differential and integral calculus – has considerable practical utility. These parts of mathematics are, on the whole, rather dull; they are just the parts which have the least aesthetic value. The 'real' mathematics of the 'real' mathematicians, the mathematics of Fermat and Euler and Gauss and Abel and Riemann, is almost wholly 'useless' (and this is as true of 'applied' as of 'pure' mathematics). It is not possible to justify the life of any genuine professional mathematician on the ground of the 'utility' of his work. (Hardy 1940: 59–60)

Overlooking his caveat (1940: 72), many have enjoyed reproducing this quotation to point out Hardy's error, that the mathematics of Fermat and Euler and Gauss and Abel and Riemann has turned out to be extremely useful, for esoteric physical theories such as string theory, but also more practically for the encryption systems which we trust keep our financial dealings secure. But this is not my concern here. I wish rather to pay attention to Hardy's use of 'real'. Elsewhere he talks in a similar vein of pieces of mathematics being 'important' and even 'serious'. I have dropped his scare quotes. It is hard to see that they can achieve very much in our times.

Hardy is being extremely exacting here on mathematicians who want to join the real mathematicians' club. I think we can afford to be considerably more generous. Where second-rate mathematicians are given short shrift by Hardy, I am willing to give even computers a fair hearing, and, although I shall not be speaking of them, people employing 'dull' calculus are not to be excluded. But that having been said, Fermat and Euler and Gauss and Abel and Riemann, along with Hilbert and Weyl and von Neumann and

Grothendieck, are right there at the core of what I am taking to be real mathematicians.

What then of the *philosophy* of real mathematics? The intention of this term is to draw a line between work informed by the concerns of mathematicians past and present and that done on the basis of at best token contact with its history or practice. For example, having learned that contemporary mathematicians can be said to be dealing with structures, your writing on structuralism without any understanding of the range of kinds of structure they study does not constitute for me philosophy of real mathematics. But, then, how exacting am *I* being?

1.2 THE CURRENT STATE OF PLAY

Ian Hacking opens his book *Representing and Intervening* with a quotation from Nietzsche's *The Twilight of the Idols*:

You ask me, which of the philosophers' traits are idiosyncracies?
For example: their lack of historical sense, their hatred of becoming, their Egypticism. They think that they show their respect for a subject when they dehistoricize it – when they turn it into a mummy.

He then continues: 'Philosophers long made a mummy of science. When they finally unwrapped the cadaver and saw the remnants of an historical process of becoming and discovering, they created for themselves a crisis of rationality. That happened around 1960' (Hacking 1983: 1).

If this portrayal of mid-twentieth century philosophy of science strikes a chord with you, you may well then ask yourself whether mathematics was faring similarly at the hands of philosophers at that time. Hacking's reference to the year 1960 alludes, of course, to the rise within philosophy of science of a movement which took the history of science as a vital fount of information, epitomised by Kuhn's *The Structure of Scientific Revolutions* (Kuhn 1962). Imre Lakatos, with his motto 'Philosophy of science without history of science is empty; history of science without philosophy of science is blind' (1978a: 102), made his own distinctive contribution to this movement. And yet, as the second epigraph of this chapter suggests, we should remember that the rationalist theory of scientific methodology he proposed and developed in the late 1960s and early 1970s derived from ideas developed in his earlier mathematical text *Proofs and Refutations*, which had appeared as a series of journal articles at around the same time as Kuhn's *Structure*. There we find sharp criticisms of a process similar to mummification, the treatment of an evolving body of knowledge as lifeless, levelled now at formalist and logicist philosophers and mathematicians:

> Nobody will doubt that some problems about a mathematical theory can only be approached after it has been formalised, just as some problems about human beings (say concerning their anatomy) can only be approached after their death. But few will infer from this that human beings are 'suitable for scientific investigation' only when they are 'presented in "dead" form', and that biological investigations are confined in consequence to the discussion of dead human beings – although, I should not be surprised if some enthusiastic pupil of Vesalius in those glory days of early anatomy, when the powerful new method of dissection emerged, had identified biology with the analysis of dead bodies. (Lakatos 1976: 3n.)

Someone working closer to the 'glory days' of early logical reductionism was Ludwig Wittgenstein. Employing imagery similar to that of Hacking and Lakatos, he writes of Russell's logicist analysis of mathematics, 'The Russellian signs veil the important forms of proof as it were to the point of unrecognizability, as when a human form is wrapped up in a lot of cloth' (Wittgenstein 1978: 162, remark III-25). But Lakatos went further than Wittgenstein in reporting to us what lay under the cloth. He exposed much more of the physiology of the mathematical life-form. So did his revelations lead to a parallel 'crisis of rationality' in the philosophy of mathematics?

To provide us with the means to gauge the situation, let us briefly sketch the current state of a central branch of philosophy of science – the philosophy of physics. Now, the first thing one notices here is the extensive treatment of recent and contemporary developments. Consider, for instance, the volume – *Physics meets Philosophy at the Planck Scale* (Callender and Huggett 2001). As this striking title suggests, philosophers of physics may interest themselves in specific areas at the forefront of physics research and yet still ask palpably philosophical questions about time, space and causation. By contrast, elsewhere one finds less specific, more allusive, studies of the way research is conducted. For instance, a book such as *Models as Mediators* (Morgan and Morrison 1999) analyses the use of models over a wide range of physics as a part of the general programme of *descriptive epistemology*. Issues here are ones just about every physicist has to deal with, not just those striving to read the mind of God. So, on the one hand, we have philosophical and historical analysis of particular physical theories and practices, while, on the other, we have broader treatments of metaphysical and epistemological concerns, grounded on detailed accounts of physicists' activities. There is a creative interaction between these two strands, both of which are supported by the study of physical theories, instrumentation and experimental methodologies of earlier times, and there is even a specialist journal – *Studies in History and Philosophy of Modern Physics* – devoted to physics after the mid-nineteenth century.

Now, certainly one can point to dissension in practitioners' visions of what philosophy of physics activity should be like. Indeed, one can construe passages of Cartwright's *The Dappled World* (1999a, see, e.g., pp. 4–5) as a call for a philosophy of real *physics*. Nevertheless, there is a strong common belief that one should not stray too far from past and present practice. How different things are in the philosophy of mathematics. While there is a considerable amount of interest in the ways mathematicians have reasoned, this is principally the case for the nineteenth century and earlier and is usually designated as *history*. By far the larger part of activity in what goes by the name *philosophy of mathematics* is dead to what mathematicians think and have thought, aside from an unbalanced interest in the 'foundational' ideas of the 1880–1930 period, yielding too often a distorted picture of that time. Among the very few single-authored works on philosophy of recent mathematics, perhaps the most prominent has been Penelope Maddy's (1997) *Naturalism in Mathematics*, a detailed means–end analysis of contemporary set theory. We shall return to Maddy's work in chapter 8, simply noting for the moment that its subject matter belongs to 'foundational' mathematics, and as such displays a tendency among practice-oriented philosophers not to stray into what we might call 'mainstream' mathematics. This tendency is evident in those chapters of *Revolutions in Mathematics* (Gillies 1992) which address the twentieth century.

The differential treatment of mathematics and physics is the result of fairly widely held beliefs current among philosophers to the effect that the study of recent mainstream mathematics is unnecessary and that studies of pre-foundational crisis mathematics are merely the historical chronicling of ideas awaiting rigorous grounding. Now, there are two ways to try to counteract such notions. First, one just goes ahead and carries out philosophical studies of the mainstream mathematics of the past seventy years. Second, one tries to confront these erroneous beliefs head on. Those who prefer the first strategy may wish to skip the next section, but anyone looking for ways to support the philosophical study of real mathematics may profit from reading it.

1.3 THE FOUNDATIONALIST FILTER

Various versions of the thought that it is right that mathematics and physics be given this very uneven treatment because of inherent differences between the disciplines have been expressed to me on several occasions when I have been proposing that philosophers could find plenty of material to mull over in post-1930 mainstream mathematics (algebraic topology, differential geometry, functional analysis, analytic number theory, graph theory, . . .). They have taken two forms:

(1) Mathematics differs from physics because of the retention through the centuries of true statements. While scientific theories are continually modified and overthrown, many true results of Euclidean geometry were correctly established over 2,000 years ago, and mankind has known arithmetic truths much longer even than this. Thus, contemporary mathematics possesses no philosophically significant feature to distinguish it from older mathematics, especially when the latter has been recast according to early twentieth-century standards of rigour. Arithmetic and its applications will provide sufficiently rich material to think through most questions in philosophy of mathematics. And even if one wished to take a Lakatosian line by analysing the *production* of mathematical knowledge and the dialectical evolution of concepts, there is no need to pick case studies from very recent times, since they will not differ qualitatively from earlier ones, but will be much harder to grasp.

(2) The mathematics relevant to foundational questions, which is all that need concern philosophers, was devised largely before 1930, and that which came later did not occur in mainstream branches of mathematics but in the foundational branches of set theory, proof theory, model theory and recursion theory. Physics, meanwhile, is still resolving its foundational issues: time, space, causality, etc.

As to point (1), I freely admit that I stand in awe of the Babylonian mathematical culture which could dream up the problem of finding the side of a square field given that eleven times its area added to seven times its side amounts to $6\frac{1}{4}$ units. Their method of solution is translatable as the calculation of what we would write

$$\{\sqrt{[(7/2)^2 + 11\cdot(6\tfrac{1}{4})]} - (7/2)\}/11 = 1/2,$$

suggesting that quadratics were solved 4,000 years ago in a very similar fashion to the way we teach our teenagers today. But, from the perspective of modern algebra and the contemporary study of algorithms, think how differently we interpret this calculation of the positive solution of a quadratic equation. As for the geometry of the Greeks, again it goes without saying an extraordinary achievement, but out of it there emerged a discipline which has undergone drastic reinterpretations over the centuries.

Today, one way mathematicians view Euclid's Elements is the study of a case of n-dimensional Euclidean geometry, the properties of the principle bundle $H \rightarrow G \rightarrow G/H$, where G is the Lie group of rigid motions of Euclidean n space, H is the subgroup of G fixing a point designated as the origin, and G/H is the left coset space. From being the geometry of the space we inhabit, it has now become just one particular species of geometry alongside non-Euclidean geometries, Riemannian geometries, Cartan geometries and, in recent decades, non-commutative and quantum geometries. Euclidean space now not only has to vie for our attention with hyperbolic space and Minkowski space, but also with q-Euclidean space. What distinguishes mathematical transformations or revolutions from their scientific counterparts is the more explicit preservation of features of earlier theories, but, as several contributors to Gillies (1992) have shown, they survive in a radically reinterpreted form. There are meaningful questions we can ask about Euclidean geometry which could not have been posed in the time of Riemann or even of Hilbert, and which would have made no sense at all to Euclid. For example, does two-dimensional Euclidean geometry emerge as the large-scale limit of a quantum geometry? The fact that we are able to ask this question today demonstrates that the relevant constellation of absolute presuppositions, scene of inquiry, disciplinary matrix, or however you wish to phrase it, has simply changed.

Moreover, to the extent that we wish to emulate Lakatos and represent the discipline of mathematics as the growth of a form of knowledge, we are duty bound to study the means of production throughout its history. There is sufficient variation in these means to warrant the study of contemporary forms. The quaint hand-crafted tools used to probe the Euler conjecture in the early part of the nineteenth century studied by Lakatos in *Proofs and Refutations* have been supplanted by the industrial-scale machinery of algebraic topology developed since the 1930s. And we find that computer algebra systems are permitting new ways of doing mathematics, as may automated theorem provers in the future. No economist would dare to suggest that there is nothing to learn from the evolution of industrial practices right up to the present, and neither should we.

An adequate response to (2) must be lengthier since it arises out of core philosophical conceptions of contemporary analytic philosophy. In the remainder of this section I shall sketch out some ideas of how to address it, but, in some sense or other, the whole book aims to tempt the reader away from such ways of thinking. Straight away, from simple inductive considerations, it should strike us as implausible that mathematicians dealing with number, function and space have produced nothing of philosophical significance in the past seventy years in view of their record over the previous three centuries. Implausible, that is, unless by some extraordinary event in the history of philosophy a way had been found to *filter*, so to speak, the findings of mathematicians working in core areas, so that even the transformations brought about by the development of category theory, which surfaced explicitly in 1940s algebraic topology, or the rise of non-commutative geometry over the past seventy years, are not deemed to merit philosophical attention. This idea of a 'filter' is precisely what is fundamental to all forms of neo-logicism. But it is an unhappy idea. Not only does the foundationalist filter fail to detect the pulse of contemporary mathematics, it also screens off the past to us as not-yet-achieved. Our job is to dismantle it, in the process demonstrating that philosophers, historians and sociologists working on pre-1900 mathematics are contributing to our understanding of mathematical thought, rather than acting as chroniclers of proto-rigorous mathematics.

Frege has, of course, long been taken as central to the construction of this foundationalist filter, but over the past few years new voices have been heard among the ranks of scholars of his work. Recent reappraisals of his writings, most notably those of Tappenden, have situated him as a *bona fide* member of the late nineteenth-century German mathematical community. As is revealed by the intellectual debt he incurred to Riemann, Dedekind and others, his concern was with the development of a foundational system intimately tied to research in central mathematical theories of the day. In this respect his writings are of a piece with the philosophical work of mathematicians such as Hilbert, Brouwer and Weyl. By contrast, in more recent times philosophers have typically chosen to examine and modify systems in which all, or the vast majority, of mathematics may be *said* to be represented, but without any real interest for possible ways in which distinctions suggested by their systems could relate to the architectural structure of the mainstream. Even distinctions such as finitary/infinitary, predicative/impredicative, below/above some point in the set theoretic hierarchy, constructive/non-constructive have lost much of their salience, the latter perhaps less so than the others.[1] How much less relevant to mathematics are the ideas of fictionalism or modalism.

A series of important articles by Tappenden (see, for example, his 1995) provides the best hope at present of bringing about a *Gestalt* switch in the

---

[1] This is largely through the reinterpretation of constructiveness by those working in computer science, but also through the desire of mathematicians to be more informative, as when a constructive proof of a result in algebraic geometry permits it to be applied to a parameterised family of entities rather than a single one. Both kinds of reinterpretation are well described by category theory.

way Frege is perceived by the philosophy community, thereby weakening the legitimising role he plays for the activity of many philosophers of mathematics. Frege should now be seen not merely as a logical reductionist, but as someone who believed his logical calculus, the *Begriffsschrift*, to be a device powerful enough to discern the truth about what concepts, such as number, are really like, sharp enough to 'carve conceptual reality at the joints' (Tappenden 1995: 449). With considerable justification Tappenden can say:

The picture of Frege which emerges contains a moral for current philosophical study of mathematics. We appear to have arrived at a stultifyingly narrow view of the scope and objectives of foundations of mathematics, a view we read back into Frege as if it could not but be Frege's own. (Tappenden 1995: 427)

For the moment, however, I choose to take a closer look at a similar reinterpretation of Frege appearing in an article written by Mark Wilson (1999), since it reveals clearly, although not altogether intentionally, the fault lines running through contemporary philosophy of mathematics. To prepare ourselves to draw some morals for our discipline from his exercise in the methodological exegesis of a hallowed ancestor it will help us to conceive of contemporary research activity in philosophy of mathematics in terms of a Wittgensteinian family resemblance. From this perspective, Wilson is aware that he is putting into question the right of a prominent clan, which includes the Neo-Fregeans, to claim exclusive rights to the patrimony of a noble forefather. Indeed, he writes 'I doubt that we should credit any Fregean authority to the less constrained ontological suggestions of a Crispin Wright' (Wilson 1999: 257). As someone who identifies with this clan ('our Frege'), he naturally finds this result unwelcome. He then continues by introducing his next paragraph as a 'happier side to our story', which oddly he concludes by indicating, in effect, that another clan – the category theorists – may now be in a stronger position to stake their claim to be seen as Frege's legatees. Interpreting this in my genealogical terms, we might say that some new shared family traits have been discovered. Just like Frege, the category theorist is interested in the organisation of basic mathematical ideas and looks to current 'mainstream' research for inspiration. In the case of Frege it was, according to Wilson, von Staudt's geometry and Dedekind's number theory,[2] while in the case of the category theorists, algebraic topology and algebraic geometry have provided much of the impetus.

---

[2] Currently, the best piece on Frege's mathematical milieu is Tappenden's unpublished 'A Reassessment of the Mathematical Roots of Frege's Logicism I: The Riemannian Context of Frege's Foundations'.

We should also note, however, that Wilson's interest in the methodological resources available to Frege and his awareness of their continued usage into more recent times is indicative of the work of yet another clan within philosophy of mathematics, the practice-oriented philosophers, or what I am calling philosophers of real mathematics. Continuing Lakatos's approach, researchers here believe that a philosophy of mathematics should concern itself with what leading mathematicians of their day have achieved, how their styles of reasoning evolve, how they justify the course along which they steer their programmes, what constitute obstacles to these programmes, how they come to view a domain as worthy of study and how their ideas shape and are shaped by the concerns of physicists and other scientists. Wilson, allied with one clan, has conducted some research in the style of a second clan, whose effect is a reduction in the legitimisation of the activities of the first clan in favour of those of a third clan.

There are traits suggesting considerable kinship between the latter two clans, the philosophers of real mathematics and the category theorists, an obvious reason for which being that category theory is used extensively in contemporary practice. Thus, the boundary between them is not at all sharp. Tappenden in his (1995) effectively casts Frege as a precursor of the former approach, but interestingly gives an example (p. 452) using category theory to illustrate how a mathematical property can be said to be mathematically valuable.

The rise of category theory will most likely be treated in different ways by the two clans: on the one hand, as the appearance, or the beginnings of the appearance, of a new foundational language; on the other hand, as an indication that mathematics never stops evolving even at its most fundamental level. In the broader context of general philosophy, the category theorist may also be led to find further roles for category theory within philosophy, for instance, to think category theory semantics should replace Tarskian set theoretic semantics in the philosophy of language (see Macnamara and Reyes 1994 and Jackendoff *et al.* 1999).

1.4 NEW DEBATES FOR THE PHILOSOPHY OF MATHEMATICS

Even were they to lose the endorsement of Frege, neo-logicist philosophers of mathematics could still claim that they are acting in accordance with current conceptions of philosophy. After all, they typically start out from the same or similar philosophical questions as those asked in philosophy of science – How should we talk about mathematical truth? Do mathematical terms or statements refer? If so, what are the referents and how do we have access to them? It just so happens, they can claim, that these questions do not lead on to further questions relevant to what takes place in mathematics departments. Where the realist beliefs of a philosopher of physics may dictate that she holds that electrons exist, but lines of magnetic force do not, or those of a philosopher of psychology that the Freudian unconscious exists, but IQ does not, mathematics treats things made of the same stuff – sets, extensions of concepts, possible constructions, fictions or whatever – so the philosopher of mathematics cannot make similar kinds of distinction.

If we pause to think about this, however, should we not consider it a little strange that whatever our 'ontological commitments' – a notion so central to contemporary English-language philosophy – *vis-à-vis* mathematics they can play no role in distinguishing between entities that receive large amounts of attention, Hopf algebras, say (see appendix), and some arbitrarily cooked up algebraic entities. If I define a *snook* to be a set with three binary, one tertiary and a couple of quatenary operations, satisfying this, that and the other equation, I may be able to demonstrate with unobjectionable logic that all finite snooks possess a certain property, and then proceed to develop snook theory right up to noetherian centralizing snook extensions. But, unless I am extraordinarily fortunate and find powerful links to other areas of mathematics, mathematicians will not think my work worth a jot. By contrast, my articles may well be in demand if I contribute to the understanding of Hopf algebras, perhaps via noetherian centralizing Hopf algebra extensions.

Surely, the philosopher ought to be able tell us something about the presuppositions operating in the mathematical community today which would account for this difference. Resorting to the property of having been used in the natural sciences will not do, since there are plenty of entities deemed crucial for the life of mathematics that have found no direct applications. On the other hand, it is hard to see how the property of being deemed thus crucial can be salient to dominant philosophical modes of thinking. For this, questions of conceptual meaning and shared understanding would have to come to centre stage. The Hopf algebra concept possesses a cluster of interrelated meanings, one of which allows for descriptions of interaction between processes of composition and decomposition in many situations. These meanings are implicated in the uses to which Hopf algebras are put.

Returning to the philosophy of science, is it the issue of realism as opposed to instrumentalism – whether we should think of unobservable theoretical entities as really existing – which can be said to relate to the most penetrating analyses of how the natural sciences work? One recent endeavour to escape the realist/instrumentalist impasse in the philosophy of science is *structural realism*, the thesis that science is uncovering only the mathematical structure inherent in the world. But the move to structural realism does not free us from having to make a stark choice as to whether mathematical entities exist or not. Indeed, the choice for the 'ontic' structural realist (see Ladyman 1998) lies between, on the one hand, some mathematical structures existing as actualised in the universe and, on the other, all mathematical structures existing, the ones we self-conscious human structures encounter being deemed physical. Now, at least, mathematicians may be said to be studying something real, rather than merely creating fictions, but still we gain no sense of mathematical thinking as part of mathematical practice. We may have been led to use specific Hopf algebras to allow us to perform calculations with Feynman diagrams (Kreimer 2000), but it cannot be right to say that they are structures instantiated in the world. Still we cannot distinguish between snooks and Hopf algebras.

An attempt to encourage the reorientation of philosophy of science towards debates better grounded in scientific practice has been made by Ian Hacking (1999). These debates are fuelled by the work emerging from science studies and sociology of scientific knowledge, which for him are 'where the action has been in the philosophy of science over the past few years' (Hacking 1999: 186). The first of the 'sticking points' on which the debates depend is related to structural realism, although without its physical foundationalism. Hacking points to an older sense of realism – the thesis that opposes nominalism – and because of the baggage associated with the term realism, he opts for the expression *inherent-structurism* (1999: 83), the position that the 'world may, of its own nature, be structured in the ways in which we describe it'. To understand what is at stake here we don't have to turn to esoteric physical theories, but rather may think through the issue by way of a question such as: To what extent is it the case that the world is structured of its own nature in such a way that it is correct to designate as 'swans' those black feathered things swimming on the Swan River in Perth, Australia, and those white feathered things swimming on the River Thames in England? Note that this is not an all or nothing kind of question. Answers will invoke ideas from anatomy, physiology, genetics, evolutionary theory, the history of ornithology, the history of colonial science, etc.

Could a parallel move work for mathematics? At first glance it might not look promising. How can we talk of a mathematical 'nature' possessing joints to carve? But this, in essence, is how many mathematicians do talk. Rather than anything contained within the doctrine currently referred to as 'Platonism', the sense they have is that something much stronger than logic offers resistance to their efforts, and that when they view matters 'correctly' things fit into place. Whereas Hopf algebra theory is an established part of real mathematics, snook theory is not, they would say, because it is not the result of carving 'conceptual reality' at the joints. This notion of conceptual reality is independent of how we might describe the nature of the stuff talked about by mathematics. It could inhabit Plato's heaven or it could be what results from the process of postulating rules or it could concern operations, actual or idealised, that we can perform on the physical world.[3]

Lakatos is aiming at Hacking's nominalist-inherent structurist distinction when he maintains that:

As far as naïve classification is concerned, nominalists are close to the truth when claiming that the only thing that polyhedra have in common is their name. But after a few centuries of proofs and refutations, as the theory of polyhedra develops, and theoretical classification replaces naïve classification, the balance changes in favour of the realist. (Lakatos 1976: 92n.)

For Lakatos, if human inquiry allows the dialectical play of ideas to occur with sufficiently little interference, it will eventually arrive at the *right* concepts. In this respect, vast tracts of logically sound, but uncritically generated, mathematics should be cast out as worthless. In response, the nominalist might say that there is nothing which intrinsically determines whether mathematical concepts have been produced correctly. What provides resistance to the mathematician are the conventions operating in her community brought about by the contingencies of history. And so we arrive at a sticking point. Out of this disagreement it might be hoped that the production of a rich picture of mathematical thinking will ensue.

Let us continue with the other two 'sticking points' Hacking sees at the heart of the science wars. These concern the inevitability or contingency of the science we have, and whether external or internal explanations should be given for the stability of our knowledge. What I find attractive about these questions is the possibility to escape polarised answers. Indeed, Hacking amusingly suggests that one locate oneself on a scale from 1 to 5. These ratings are presented in absolute terms as though we have to give a single answer to, say, how likely we reckon it is for specific scientific developments to have occurred. It seems to me more reasonable to take it as a measure of the tendency within one to take a certain side in a series of arguments. We all know of colleagues who tend to take up more contingentist or necessitarian views than ourselves on just about any question.

---

[3] These last two are, of course, distinguishable: you can physically move a knight forward one square, but the rules of chess do not allow you to do so.

Each of these additional sticking points is relevant to mathematics in the sense that we may argue about the following kinds of question: Is it the case that had a successful mathematical discipline been developed to a level of sophistication comparable to our own, then it would have to involve something equivalent to X, where for X we may substitute the natural numbers, the rationals, the complex numbers, complex analysis, Riemann surfaces, finite groups, Lie groups, Hopf algebras, braided monoidal bicategories, etc.? Why do we still adhere to, and teach undergraduates about, certain ways of thinking of X?[4]

We can find examples of these debates already happening. Indeed, on the question of contingency, Lakatos and Bloor use the same material, Lakatos's case study of the Euler conjecture from *Proofs and Refutations*, to argue different sides. Lakatos tells us that:

any mathematician, if he has talent, spark, genius, communicates with, feels the sweep of, and obeys this dialectic of ideas. (Lakatos 1976: 146)

While for Bloor:

Lakatos's discussion of Euler's theorem . . . shows that people are not governed by their ideas or concepts . . . it is people who govern ideas not ideas which control people. (Bloor 1976: 155)

Now, to Hacking's trio of sticking points I would like to add two more. First, there is the issue of the unity or connectivity of mathematics. This is nothing to do with all mathematical entities being seen as constructible within set theory, but much to do with cases of unexpected discovery such as finding that when using Hopf algebras to calculate expansions in perturbative quantum field theory, answers depend on values of the Riemann zeta function. There is an inclination to rebel against such a story and so to latch on to an image of mathematics as thoroughly fragmented as Mehrtens (1990) chooses to do, but then we need explanations of cases of surprising connectivity. For instance, how is it that a geometry devised after a failed *reductio ad absurdum* argument, starting out from the negation of Euclid's fifth postulate, could provide a useful classifier in knot theory in that it allows for the measurement of the volume of the hyperbolic space that typically remains when a knot is removed from the space in which it sits? For those who admit a considerable degree of unity, the further

---

[4] For an attempt to answer the question 'What kind of combination between the "natural" and the historically contingent led to our conception of modern logic?' by arguing that first-order logic is 'no natural unity' see Ferreirós (2001).

question arises of its causes: social pressures to keep to certain ways of thinking, the way our brains work, or encounters with inherent structure.

Second, there is the issue of the explicability of the applicability of mathematics. Usually this is polarised into 'it's an inexplicable miracle how mathematics, developed for aesthetic reasons, applies to the world' position opposed to one asserting 'it's not surprising because mathematics has been thoroughly shaped by the concerns of physicists'. Think how much more we might learn from a debate between, on the one hand, someone at point 3 on the scale, who recognises mathematics as arising from what the world allows us to do it, and who knows how intricately linked mathematics and physics were in the nineteenth century, but who still thinks there is something to explain about how Riemannian geometry was there for Einstein, and on the other hand, someone at point 4 who reckons in addition that physicists configure their theories to allow for the use of available mathematics. Mark Steiner (1998) has provided a start for us, but there are many more subtleties to discover. Just read a mathematician on the subject to feel the contemporary richness of this issue (e.g. Klainerman 2000).

These debates are not just about getting our description of mathematical practice right, but bear on ideas about how things ought to be. Just as there is a normative element to Lakatos's remarks about realism – we ought to follow his methodology to arrive at 'real' classifications, with the suggestion that we may, and indeed often do, fail to do so – so each of the other sticking points can be made to bear some normative load. For instance, we hear that mathematics may be fragmented today, but along with physics, it could and should be unified by adopting the language of geometric calculus (Hestenes 1986).

These kinds of questioning are to be addressed by an understanding of mathematical knowledge as historically situated rather than timeless. Lakatos understood this, but his work was only a start. To move on we shall need a revolution of sorts. In the 1960s Kuhn was able to revolutionise the philosophy of science partly because there was already a considerable body of history and sociology of science in existence, the product of professionalised disciplines. Philosophy of physics was already a much larger affair than its mathematical counterpart, with ahistoricist philosophers well grounded in mainstream theories and experiments connected with general relativity and quantum mechanics. We should remember, for instance, that Reichenbach worked for a time with Einstein. On the other hand, the logicism expounded by Reichenbach, Hempel and others of that generation was too deeply ingrained in the philosophical psyche to be overcome easily. By the 1960s, there was no philosophical tradition requiring extensive mathematical knowledge, and the history of modern mathematics was still largely an amateur affair stuck at the stage of 'Men of Mathematics', and so the conditions were not right for *Proofs and Refutations* to have its effect.

Forty years on, few philosophers of mathematics have been prompted to gain anything approaching the level of historical and theoretical knowledge that philosophers of natural science are expected to have. This is partly owing to the state of the history of mathematics. We still have nothing to compare with the sophistication of contemporary history of modern physics, the history of twentieth-century mathematics remaining largely the preserve of mathematicians. But these factors would be of little importance were the philosophical agenda to require serious engagement with the thinking of mathematicians through the ages.

How radical a change is required? It often seems that anyone wishing to take the history of a science seriously in their philosophy requires what to many in the English-speaking world of philosophy is an unorthodox philosophical background. This Lakatos certainly had. For Kuhn, on the other hand, it was implicitly fed to him via the historians he studied, Koyré, etc.:

the early models of the sort of history that has so influenced me and my *historical* colleagues is the product of a post-Kantian European tradition which I and my *philosophical* colleagues continue to find opaque. Increasingly, I suspect that anyone who believes history may have a deep philosophical import will have to learn to bridge the longstanding divide between the Continental and English-language philosophical traditions. (Kuhn 1977: xv)

Without the resources of a dialectical philosophy, Kuhn came unstuck. In the rigid epistemological framework he inherited from the logical empiricists, sameness and difference were polarised, a concept could not evolve into another while retaining something of its past. And so he was guilty both of underestimating diversity within a paradigm and of overestimating incommensurability between paradigms.

One of the last of the English-language philosophers not to be cut off from Continental thinking by the rising tide of analytic philosophy was R. G. Collingwood. Collingwood had the notion that a discipline in any particular epoch possesses its own constellation of absolute presuppositions, and that discovering these is the task of the metaphysician. The fact that these absolute presuppositions change is sometimes seen as having as its consequence that there exists between the stages of development of a discipline an incommensurability akin to Kuhn's. This, however, is a misunderstanding of Collingwood's position.[5] Aside from the possibility of there being absolute presuppositions which have been maintained since the Greeks, when change does takes place it need not be construed as a discontinuous rupture, but rather as a dialectical change in which something about the earlier presupposition is retained in whatever it turns into:

The problem of knowledge is therefore everywhere and always the same in its general form: when we are presented with something which we do not understand . . . we are to reach an understanding of it by finding out how it has come to be what it is: that is to say, by learning its history. (Collingwood 1999: 178)

This kind of understanding of change was part and parcel of Lakatos's thinking, as his desire to become the founder of a dialectical school in the philosophy of mathematics reveals (Larvor 1998: 9).

For Collingwood, along with this dialectical sensitivity, a capacity to experience the force of the absolute presuppositions of the contemporary form of the discipline about which one is philosophising is vital. While describing which qualities someone should possess to be able to answer the questions of philosophy of history, he remarks acidly that:

No one, for example, is likely to answer them worse than an Oxford philosopher, who, having read Greats in his youth, was once a student of history and thinks that this youthful experience of historical thinking entitles him to say what history is, what it is about, how it proceeds, and what it is for. (Collingwood 1946: 8)

A similar conclusion could be formulated for philosophy of mathematics, and indeed Kant is praised for dealing with the presuppositions of mathematics 'rather briefly' for 'he was not very much of a mathematician; and no philosopher can acquit himself with credit in philosophizing at length about a region of experience in which he is not very thoroughly at home' (Collingwood 1940: 240).[6] Returning to history, he continues:

An historian who has never worked much at philosophy will probably answer our four questions in a more intelligent and valuable way than a philosopher who has never worked much at history. (Collingwood 1946: 9)

Evidence for the equivalent statement about mathematics is provided by the very many important contributions made by mathematicians thinking about their discipline, several of which I shall lean on in the course of this

---

[5] See Oldfield (1995) on this point.

[6] Collingwood is being rather unfair to Kant in that, as Friedman (1992) argues, Kant's engagement with mathematics and especially physics was what gave depth to his philosophy. But that then only supports Collingwood's thesis that to do philosophy of a discipline well one must be 'thoroughly at home' with it.

book. These include the thoughts of Weyl, Weil, Mac Lane, Rota, Atiyah, and from the current generation, Gowers and Baez.

1.5 TOWARDS A PHILOSOPHY OF REAL MATHEMATICS

Aspray and Kitcher (1988: 17) dub as belonging to the 'Maverick Tradition' those philosophers of mathematics who pose such questions as:

> How does mathematical knowledge grow? What is mathematical progress? What makes some mathematical ideas (or theories) better than others? What is mathematical explanation?

While their portrayal of such philosophers as non-conformists may not be far off the mark, it clearly does not represent a desirable state of affairs. Language has a performative role as well as a descriptive one, and we should be looking to inspire a new generation of philosophers to sign up to the major project of understanding how mathematics works. Maddy has opted with her *naturalist methodology* not to use the word 'philosophy', which seems to me an unnecessary concession. Larvor (2001) has described a movement he terms the *dialectical philosophy of mathematics*, and kindly refers to me as one of its three leading exponents. Then again my *philosophy of real mathematics* may provide a louder clarion call for a time.

One way to proceed with this programme is to return to two of the founding fathers of the philosophy of real mathematics: Pólya and Lakatos. This I shall do, but in full consciousness of a problem we face. Back in the early 1960s, Lakatos and Kuhn were able to take risks with their pioneering historicist philosophies of mathematics and science, where bold theses were defended on the basis of a handful of sketchy historical reconstructions. Now, from the perspective of our current sophisticated science studies we look back on Kuhn's *Structure* as being rather simplistic, if understandably so, and we may agree with Peter Galison (1997) that it would be extremely naïve today to maintain that there is a unique structure to scientific revolutions. Our discipline has not had the same opportunity to grow up, and so forty years on we find ourselves in an awkward situation. We wish to propose striking theses, since tentatively expressed claims are hardly likely to energise our field, and yet with so little to build on it is likely that our efforts will appear immature by comparison to our sister discipline. For instance, historians, philosophers and sociologists of science may wonder whether it is necessary to rake up all the paraphernalia of Lakatos's *research programmes*, as I do in chapter 7, when they now have little time for them.

And isn't the Bayesianism of chapters 4 and 5 beyond the pale? Hopefully, they will allow us a period for recapitulation.

An alternative strategy would be to claim sanctuary under the protection of philosophy of science on the pretext that mathematics be seen as a science. Now, one of the varieties of disunity treated by Ian Hacking in his paper 'The Disunities of the Sciences' (Hacking 1996) he terms *methodological* disunity, which concerns the diversity of styles of scientific activity. For several years he has expressed support for the following classification of scientific styles proposed by the historian of science A. C. Crombie:

> (a) postulation in the axiomatic mathematical sciences, (b) experimental exploration and measurement of complex detectable relations, (c) hypothetical modelling, (d) ordering of variety by comparison and taxonomy, (e) statistical analysis of populations, and (f) historical derivation of genetic development. (Hacking 1996: 65)

To these Hacking wishes to add 'laboratory science . . . characterized by the construction of apparatus intended to isolate and purify existing phenomena and to create new ones' (*ibid.*). Hacking applauds Crombie's inclusion of (a) as 'restoring mathematics to the sciences' (*ibid.*) after the logical positivists' separation, and extends the number of its styles to two by admitting the algorithmic style of Indian and Arabic mathematics. I am happy with this line of argument, especially if it prevents mathematics being seen as activity totally unlike any other. Indeed, mathematicians do more than postulate axioms and devise algorithms; it would hardly be figurative to say that mathematicians also engage in styles (b) (see chapter 3), (c) and (d),[7] and along the lines of (e) mathematicians are currently analysing the statistics of the zeros of the Riemann zeta function.[8] As for Hacking's additional scientific style – the construction of apparatus – Jean-Pierre Marquis (1997) made a start on analysing the notion that some mathematical constructions are used as machinery or apparatus to explore the features of other

---

[7] Cf. John Thompson's comments: 'the classification of finite simple groups is an exercise in taxonomy. This is obvious to the expert and to the uninitiated alike. To be sure, the exercise is of colossal length, but length is a concomitant of taxonomy. Those of us who have been engaged in this work are the intellectual confreres of Linnaeus. Not surprisingly, I wonder if a future Darwin will conceptualize and unify our hard won theorems. The great sticking point, though there are several, concerns the sporadic groups. I find it aesthetically repugnant to accept that these groups are mere anomalies . . . Possibly . . . *The Origin of Groups* remains to be written, along lines foreign to those of Linnean outlook' (quoted in Solomon 2001, 345).

[8] Hacking (1992: 5) remarks that 'A great many inquiries use several styles. The fifth, statistical, style for example is now used, in various guises, in every kind of investigation, including some branches of pure mathematics.'

mathematical entities, as when, for instance, K-theory was constructed to probe topological spaces. But let us remember that these styles of mathematical activity arise in particular epochs and evolve over the centuries. After all, Hilbert's use of axiomatisation differs quite considerably from Euclid's.

The fact that there is such a degree of overlap between the styles of mathematical and scientific activity suggests we might learn from current studies of scientific argumentation. However, were we to ignore the differences between, say, classifying finite simple groups and tabulating their properties, and doing similarly for the chemical elements, fundamental particles, or zoological phyla, we would lose what is unique about mathematics. For one thing, these styles of activity work in a more interactive fashion for mathematics, owing to the greater homogeneity of mathematical material. Pieces of mathematical machinery, such as homology and cohomology theories, although used as 'black boxes' by some consumers, are themselves mathematical entities and so the possible subject matter for mathematical classification, as for instance when the so-called *spectra* representing extraordinary cohomology theories are gathered together to form a category, and one of them – the sphere spectrum – shown to be maximally difficult to compute with. Of course, there are theories of instrumentation in the natural sciences, but nobody seriously contemplates the space of all possible machines of a certain kind.

As I have said, I see no intrinsic reason why we should not succeed in drawing connections between developments in mathematics, including those which have occurred in recent decades, and recognisably philosophical concerns. Indeed, we can point to a considerable number of important studies already in existence as evidence, the vast majority in the mould of descriptive epistemology. But to emulate philosophy of physics we need to make a more systematic effort to engineer space for ourselves to work with a wide range of issues. Alongside descriptions of how research mathematicians have worked, we should also allow philosophy to treat interpretational issues interior to branches of mathematics in such a way as to provide us with insight into reasonably large portions of mathematics, on the assumption that we will miss something important if we only look for features relevant to mathematics as a whole.

Not only do we need to free ourselves from the requirement that we treat simultaneously all of the space of mathematics, we also need to work out varied ways to liberate ourselves from the appeal of timelessness. In doing so temporality needs to be introduced at many scales, since mathematics more than any other discipline has been formed by the endless reinterpretation of its own results. The sociologist of science Andrew Pickering extends his 'Mangle of Practice' to mathematics in Pickering (1997). In a similar vein, and in the spirit of Diderot's *bon mot* quoted at the beginning of this chapter, we might liken doing mathematics to kneading dough. If your time-scale allowed you only to describe the lump being flattened and stretched, you would miss the coming together of widely separated points during the folding process. As a piece is torn off and set to one side, you might believe it looked finished. But had you waited a little longer you would have seen it return under the knuckle to be reshaped into the rest of the dough. No mathematical concept has reached a definitive form – everything is open for reinterpretation, even the integers as we shall see in chapter 10. There is a danger then in relying too heavily on 'one-pass history', as it is known. Pickering's story ends with Gibbs and Heaviside laying out 'the fundamentals of vector analysis, dismembering the quaternion system into more useful parts in the process' (1997: 60). One might easily draw the moral from the apparently ephemeral success of the quaternions that much was contingent about their evolution, but we can support the inevitabilist position somewhat by a longer-term history which shows that since 1880 the quaternions have re-established themselves as very respectable citizens of the world of mathematics for their many good works, and that while Pickering's claim that vector analysis is 'central to modern physics' (1997: 45) is questionable, a similar claim about the unit quaternions in the shape of the Lie group SU(2) is not.[9]

The need to avoid confining oneself to a particular historical moment is made more pressing by the fact that the material basis of mathematical practice does not anchor it so precisely in time or space as does the physicist's apparatus. Of course, this freedom is not limitless. Restrictions on the availability of papyrus or supercomputers, or of access to books and journals certainly play their part, but there is nothing quite to compare with the localised knowledge of glass production or lens grinding, or the fixity of a particle accelerator. Today, software and electronic journals can travel across the world in seconds, mathematicians in hours.

We are faced with an enormous and daunting choice. A fascinating, but extremely challenging, case to treat would be the vast ongoing programme

---

[9] Baez (2002: 146): 'The unit quaternions form the group SU(2), which is the double cover of the rotation group SO(3). This makes them nicely suited to the study of rotations and angular momentum, particularly in the context of quantum mechanics. These days we regard this phenomenon as a special case of Clifford algebras. Most of us no longer attribute the cosmic significance that Hamilton claimed for them, but they fit nicely into our understanding of the scheme of things.'

to form a non-commutative version of geometry, which among other things aims to allow non-commutative algebras of quantum mechanical observables to be thought of as algebras of functions on phase spaces. The related 'q-disease', whose primary symptom is a preference to study deformations of familiar objects rather than the originals, has by now become an epidemic. Alternatively, we find that mathematicians are especially interested in one of the debates I suggested in the previous section, the one centred on the issue of unity. Arnold (2000), for instance, tells us that many of the unexpected connections found between apparently distant regions of mathematics can be viewed as one construction, being the complexification, quaternionisation, symplectisation, or contactisation of another, and outlines how he sees large tracts of mathematics in this light. Related to this is the appearance in the post-Second World War era of a sharp divide between the highly interconnected nexus of theories such as algebraic geometry and differential topology, on the one hand, and the more fragmented treatment of partial differential equations, on the other, belied from time to time by the construction of bridges from specific kinds of differential equation to parts of the nexus.

Another topic which needs to be explored is the introduction of probabilistic and stochastic thinking into mathematics. Besides its use in graph theory, for instance in establishing lower bounds for Ramsey numbers, mathematicians have come across the surprising fact that the distribution of the zeros of the Riemann zeta function shares much in common with the distribution of energy levels of a heavy nucleus. Both are connected to the distribution of eigenvalues of very large random matrices.

A final example, one I shall be treating in this book, is the rise of a discipline known as *higher-dimensional algebra*, which aims to permit composition to take place in additional dimensions to the standard linear one of the printed text, blurring the boundary between algebra and topology in the process. As we shall see later, higher-dimensional algebra offers us an account of why Hopf algebras are 'good' things.

1.6 RELATIONS WITH THE PHILOSOPHY OF PHYSICS

Evidently, the relationship with the philosophy of physics is likely to be closest, although there is no reason why there should not be important collaborations with, say, the philosophy of biology, and we very much need to bolster the philosophy of theoretical computer science. Philosophers of physics are already in a much better position to provide us with some answers to the question 'Why Hopf algebras?', since they can point to applications in integrable models of statistical mechanics, inverse scattering, renormalisation theory, etc. But we should remember that mathematics considers more varieties of structure than does theoretical physics, even when we take the latter to include the construction of toy models. More importantly, it considers them in different ways.

I am hopeful that the philosophy of physics may take a mathematical turn and so require the services of a philosophy of real mathematics. Just as philosophers there had to overcome various logical empiricist tendencies – the emphasis placed on theory rather than experiment, modelling being taken in the logicians' sense, etc. – it is time to overcome the logicist idea that mathematics contains nothing beyond an elaboration of the consequences of sets of axioms. Until philosophy of science became serious about the actual workings of science, one could take the logical empiricist line on confirmation without worrying much about the way experiments are conducted. What seems to have changed little in this more realistic climate, however, is the tendency to take the mathematics used in physics as read, either as devised on the fly by physicists or else as selected off the shelves of the mathematicians' superstore. The conceptual contribution of mathematics to the work of physics is thus largely being overlooked. In a sense we need to mimic at the level of philosophy the recent rapprochement between mathematics and physics. A simplistic story has it that after centuries of close union, the two disciplines became divorced in the 1930s over a quarrel about quantum field theory. Each went off on its own path, and it was only in the late 1970s they discovered that they had both been performing similar constructions, e.g., connections on fibre bundles and gauge potentials. The danger in physicists ignoring the ideas of mathematicians lies in their overlooking the power of mathematical conceptualisations when they are interested in a piece of mathematics as a ready-to-wear item rather than as part of a network of associated ideas. There is a many–many relationship between mathematics and physics: one piece of physics calls upon varied pieces of mathematics, one piece of mathematics can be used in very varied situations (in computer science and mathematics itself as well). What could potentially provide some unity to the many uses of a piece of mathematics lies at the conceptual level. Similarly, mathematicians may not understand how field theoretic considerations mean that different pieces of mathematics are related.

Take our running example of Hopf algebras, a small fragment of the picture of whose uses I have drawn below:

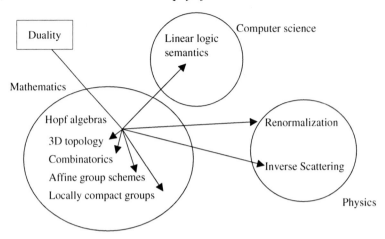

We can see an example of how mathematical understanding might impinge on physical conceptualisation in the work of Shahn Majid (1995). Associated with Hopf algebras, he tells us, is a notion of duality realised as a form of input–output symmetry. When these algebras are used as physical models, instead of taking the observables as forming an algebra and the states of the system as forming merely a set or space, one now has a multiplication imposed on the states as well. So one can reverse the interpretation of the model, interchanging states and observables and causing random walks to be seen as dual to creation processes. This may sound like a physical argument, but the idea of viewing a pairing between a function and the value of an argument both as the application of the function to the argument and as the evaluation of the function by the argument is a fundamentally *mathematical* thing to do.

The essential point here is that, despite the rapprochement between mathematics and physics, there will be no simple unification of disciplines. The mathematical physicist Cumrun Vafa expresses this well:

It is clear to many physicists and mathematicians that we are going to continue witnessing many exciting domains of interaction between physics and mathematics. So much so that some researchers are even predicting a merger between the two fields in the near future. However, I believe merging of the two subjects will not take place given the vast differences in aims that mathematicians and physicists have about a given subject. In fact, if a merger were to happen, I believe it would not even be a healthy development. Physicists and mathematicians have benefited from each other so much precisely because they have such clearly *distinct* views about a given subject. (Vafa 1999: 327)

This suggests the need for philosophers of physics and of mathematics to talk to one another, while maintaining their own interests. If philosophers of physics are to study the practice of physics, they must take into account the contemporary form of the interaction with mathematics, and vice versa.

The history of this intimate relationship between mathematics and physics has certainly left its mark, but we may attribute the shaping of contemporary mathematics to more than this. We shall now equip ourselves with the means to find our bearings in the space of ways of thinking about mathematics by considering factors shaping the way it proceeds.

1.7 FACTORS GOVERNING THE WAY MATHEMATICS PROCEEDS

We can usefully think of the path taken by mathematics through its history as governed by factors arising from a variety of sources. To give a flavour of the kinds of factors we shall meet weaving their way through later chapters, I shall treat them in groups according to the following classification: (a) logical and calculational correctness; (b) plausibility; (c) psychological factors; (d) technological factors; (e) sociological and institutional factors; (f) relations with other sciences; and (g) inherent structure. Let us consider them in turn.

(a) *Correctness within some existing calculus*: generally speaking, algebraic and analytic calculations are performed within some specified calculus, and proofs conform to the accepted logical standards of the day. It may seem unusual to group these together, since philosophers usually like to take their preferred form of logical calculus as fundamental. I tend to side, however, with Poincaré and Hilbert in viewing the structure presupposed by even the predicate calculus as substantial. In any event, it is not the concern of this book to determine how large a part of mathematical knowledge may be said to be *a priori* knowledge of analytic truths, as the neo-Fregeans imagine themselves to be doing by casting arithmetic in terms of second-order logic. What I *am* concerned with is the notion that a calculus may 'run along the grain' of a piece of mathematics. It is true that although a mathematical proof may involve some amount of algebraic or analytic calculation, it could in principle be couched purely in a logical calculus, but it is important to realise just how unappealingly long-winded the result would be. Thinking about this situation from the opposite direction, without the presence of a suitable algebraic or analytic system it is extremely unlikely that one would hit upon the proof. If these points are not obvious, the reader is encouraged to translate a simple piece of algebraic manipulation, the expansion of a power of a polynomial, say, into first-order logic. Four lines suddenly become a hundred. Now, in *Proofs and Refutations* a considerable part of the problem for the students is that they lack adequate means to define what they mean by a polyhedron. These days mathematicians have the resources on hand to make definitions precise and to present their arguments with a standardised level of rigour. This, however, is generally not sufficient. What one generally requires is a good algebraic or analytic framework.

It has proved to be the case that much is gained by constructing a formal calculus in which calculations and arguments may be said to be conducted correctly. This is as much for the provision of a boundary which shows explicitly when one is stepping outside the system, as for assuring the security of results demonstrated within it. The persistent need to breach the constraints of the system will suggest the necessity of augmenting or reconstructing the system. An early example was the study of divergent series (see Kline 1972, ch. 47). Series divergent according to Cauchy's convergence criterion were known to be useful, but it took a concerted effort for several decades after 1880 to establish the range of validity of the use of such series. More recently, we find constructions in quantum field theory causing mathematicians a headache. Take for instance Edward Witten's response to the challenge Michael Atiyah proposed in the late 1980s to find an intrinsically three-dimensional interpretation of the Jones polynomial, a recently discovered knot invariant calculated via knot projections in the plane. Witten provided a quantum field theoretic interpretation involving integration over a space of connections on a principal bundle modulo gauge equivalence. However, no appropriate measure has been defined over this infinite dimensional space, and so Witten's formula is treated as a heuristic device. Calculations may be performed, but the boundaries of what is permitted are not clear. Few mathematicians would not be delighted to see this work given a firmer grounding.

(b) *Plausibility*: an understanding of how practitioners of a discipline reason plausibly should be hugely important for philosophers of that discipline. They have tended, however, to steer clear of this topic and so we must rely on the writings of mathematicians such as George Pólya. Pólya taught us that mathematicians spend much of their time formulating conjectures, providing evidence for these conjectures by way of calculating particular instances, or drawing analogies to similar problems, and then planning proofs based on the existing knowledge and their understanding of the proofs of related results. Much of this activity has an inductive flavour to it. We need to understand how considerations of plausibility govern the choices of mathematicians at different time-scales, from their day-to-day reasoning on a specific problem to the decision to dedicate a considerable part of their lives to a particular research programme.

Consideration of which results are likely to be correct and which are likely to be accessible using available proof techniques are clearly relevant to rational decision making as to the choice of research topic. Furthermore, it is common practice to build a considerable body of mathematics upon unproven conjectures. For example, we have results which start 'If the Riemann hypothesis is correct, then . . .'. Obviously, building a pyramid of conjectures on a base of uncertain propositions would be an unwise thing to do, if we had strong grounds to doubt their truth.

(c) *Psychological factors*: the extent to which mathematics reflect the workings of the human mind, both in its strengths and limitations, is a question largely discussed by cognitive psychologists and philosophers interested in their work (cf. Kitcher 1983). Unconstrained by the need to demonstrate direct relevance to the world, it is easy to imagine that mathematicians would unintentionally project into their work features of the ways in which their minds work. Such projection would be expected to reveal itself especially when cognitive obstacles were being overcome. For instance, mathematicians in the past century and a half took a long time to become happy with the idea that a manifold should be defined as a patchwork of pieces of space held together by some specified variety of 'glue', rather than as sitting in an ambient space, as our untutored intuitions would suggest. An individual will have to surmount this hurdle through her own education. But it is possible that many of the ways psychological factors structure mathematics go unnoticed. If so, it has an important bearing on philosophy of science. The more you see our mathematics as significantly shaped by the way we think, the more you seem to be forced to find its application in the world a sign that something akin to Kantianism is correct, unless you prefer the notion of a 'user friendly' universe which our minds are somehow adapted to understand (Steiner 1998: 8).

Even if we could establish that many pieces of mathematical research bear the stamp of some universal features of mental functioning, this would not make it irrational to pursue them. If our aim is to produce more mathematics of relevance to our sciences, and it can be argued that working in a way that suits our minds is most likely to achieve this, then it is rational to do so. For instance, one may reasonably attribute psychological factors to the fact that the study of knots, that is, circles embedded in 3-dimensional space, began long before the study of knotted spheres in 4-dimensional space. But even if it were deemed at some stage very likely that the latter would turn out to be much more important in terms of its applications, perhaps as the basis of a 4-dimension quantum field theory as some practitioners of quantum gravity hope (see chapter 10), in view of our greater capacity to visualise knots and so produce an algebraic treatment of them which is likely to be relevant to the next higher dimension, it may be argued that it was rational to devote our resources to the former first. Two mathematicians working on precisely this project see the situation as follows:

> The purpose of this book is to develop the diagrammatic theory of knotted surfaces in 4-dimensional space in analogy with the classical theory of knotted and linked circles in 3-space. This goal may sound unachievable to some readers, how can we perceive phenomena that occur in 4-space? (Carter and Saito 1998: ix)

They proceed to develop *movies*, sequences of slices through a knotted surface, all of which would have been unthinkable without the help of earlier experience on knots.

(d) *Technological factors*: these include the means to communicate and publish, and the means to compute. The former are more important than one might imagine. Reviel Netz in *The Shaping of Deduction in Greek Mathematics* (Netz 1999) writes on the great expansion of mathematical activity that occurred when mathematics freed itself from its oral tradition and storing and retrieval became much more written. Much closer to our times, when textbooks had to be typeset the expense to be incurred in their production forced publishers to guarantee that they would find a substantial readership and so be cautious. With the advent of Xerox machines, ideas at the frontiers of research could be conveyed rapidly around the world. However, when you see the ghastly typewritten monographs of the 1970s, you can appreciate the limitations of the available notational devices. Now we have the opportunity to read for free papers submitted to preprint servers, written with the notational resources of Latex, and illustrated by diagrams constructed using clever software packages, and we are able to watch remotely lectures given by illustrious mathematicians and view moving graphics. But might Latex itself act as a constraint on the development of new notational practices? It is not hard to argue, then, that the lack of technological resources may delay the development of a field.

As we shall see in chapter 10, it has been claimed that it is the availability of more sophisticated printing technology in the past few years which has allowed the emergence of higher-dimensional algebra.

As for computation, I shall largely be concentrating on automated theorem proving and conjecture formation in this book, but it is clear that numerical and algebraic calculation has had the more profound effect upon the practice of mathematical research. Whole new fields of mathematics have been opened up by the increasing power of computational devices. But this gives rise to the question how the mathematical community ought to use and develop technology. Ursula Martin, a researcher working in the field of computer algebra systems, has suggested on the basis of her analysis of mathematical practice (Martin 1999) that there is little purpose to the activity of one sector of the automated theorem proving community. Members of this group seek to use automated proof assistants to produce and check completely formally correct proofs of large tracts of mathematics. Martin argues, convincingly to my mind, that this is largely a waste of effort. How much more profitable for someone to collect, and make available on the Internet, informal insights? Listen to the mathematician Raoul Bott:

With one offhand remark we give away our insight of years of thinking, and such a remark might illuminate a whole field or fit into one's brain just right to unlock some new insight. (Jackson 2001: 381)

(e) *Sociological and institutional factors*: The effects of the war effort on American mathematics in the post-Second World War era are very evident, as are the effects of the prevailing political climate of the Cold War in the Soviet Union through the 1960s and 1970s. We are told, for instance, that Russian algebraic topologists were thought rather frivolous for working in such an impractical domain, until the announcement of the Atiyah–Singer Index theorem made it evident that algebraic topology was relevant to the solutions of partial differential equations, and so its study of potential benefit in sustaining the Revolution. However, in the course of this book I shall not be dwelling on how sociological factors at the external end of the range dictate the way mathematics is done, but rather more internalist considerations of decision making within the community. This is not an indication that I underestimate the importance of the bearing of the larger society on the mathematics it supports. It has already been the subject of extensive studies (e.g. Restivo 1992), which tend to map out the history of mathematics at a large time-scale, requiring a severe compression of the twentieth century.

The arrival of improved means of communication which I mentioned in (d) above may not be all for the good. Indeed, it has been argued that they have a tendency to produce small communities of researchers linked not by spatial proximity, but solely by their narrow field of interest, thereby preventing cross-fertilisation with other branches of mathematics, and producing strong tendencies to follow the latest trends in a field. Connes, on the other hand, presents a much more optimistic picture of how the mathematics community copes with its vast expansion (Connes *et al.* 2000: 100–1).

(f) *Relations with other sciences*: the structures encountered in other disciplines, most especially physics, are often rich sources of inspiration to mathematicians. In our quest to represent the workings of mathematics we shall have to figure out the extent to which mathematics has been shaped by its relationship with the natural sciences, and how the mathematical requirements of the biological and neural sciences are currently leading mathematics in new directions.

It was famously the concern of John von Neumann that mathematics tends to advance along the path of least resistance into very specialised branches unless is it brought back periodically into contact with its empirical roots. If so, mathematicians ought to keep an eye on adjacent sciences. On the other hand, one may also argue that it is extremely likely that many mathematical constructions, even those far removed from any empirical source, have considerable unrealised potential to be applied. Who would have dreamed that our knowledge of elliptic curves could provide the basis for an encryption system? Either way few would disagree that mathematicians and scientists should talk more to each other.

(g) *Inherent structure*: one could imagine a philosopher responding to the classification up to this point as showing that everything is merely logic plus sociology. Someone from the Strong Programme might go a step further and say that our use of logic also has sociological determinants (see Barnes *et al.* 1996, ch. 7) – nothing is free of sociology. The idea of inherent structure, which may be chimerical, answers to the sense that mathematics offers resistance to the mathematician well beyond that attributable to correctness within some universal calculus, and yet not just emerging from disciplinary training. For example, there are any number of ways to generalise or deform the concept of a group perfectly rigorously, but only a very few have important properties. Again, in the late nineteenth century Dedekind and Weber postulated a strong analogy between number fields and function fields. The playing out of this analogy has formed a major part of mathematics ever since. Does its success suggest the existence of objectively similar characteristics? Can any two concepts be made to resemble each other considerably if worked hard enough?

This form of realism, resembling Hacking's notion for the natural sciences discussed in section 1.4, is independent of the issue of whether mathematical entities really exist in some ontological realm or other. It is about seeing things 'correctly', that is, how things 'really' are. This is the type of realism which underlies the suggestion that a mathematician may have 'glimpsed' something decades before another brings it into sharp focus. The key words here are 'natural' and 'fundamental'. Mathematicians use them all the time to describe features of their work that appear to arise from the nature of the domain they are studying, rather than being externally imposed.

While set theory displays certain 'foundational' virtues, we must recognise that reformulating a piece of mathematics that way may run against its 'grain'. While commenting on the success of the set theoretic programme to comprehend all mathematical entities in its terms, the mathematical physicist John Baez remarks, 'one must bend over backwards to think of such varied entities as sets, so this formalization may seem almost deliberately perverse' (Baez 2001a: 189). The price to be paid for universality is unnaturalness. Instead of seeing mathematical entities and constructions merely as ultimately composed of set theoretic dust, we should take into account structural considerations, rather as the student of anatomy gains little by viewing the human skeleton merely as a deposit of calcium. Such considerations can hardly fail to take into account what is special about those disciplines described as 'geometric' and should lead us to form a new philosophy of geometry which treats: fibre bundles, connections, curvature, spin networks and foams, non-commutative spaces, toposes, instantons, sheaves, gerbes, symplectic geometry, and geometric quantisation.

Tracing out the interrelations between these factors should prove a challenging task. My response to this challenge will run as follows.

1.8 OUTLINE OF THE BOOK

One approach to thinking about mathematics is to consider how differently it might have turned out. This may be done by engaging in counterfactual history, where, say, we imagine that Chinese mathematics had prevailed over Greek mathematics. However, the advent of the computer provides us with a more convincing strategy since machines are today starting to contribute to mathematical research in their own way. We begin part I with two chapters (chapters 2 and 3) examining the potential role of computers in mathematics. One of the lessons we learnt from *Proofs and Refutations* is that when humans do mathematics, three components of their activity – conceptualisation, the formation of conjectures and the construction of proofs – are inextricably linked. Given the very distant prospect of computers being able to supersede mathematicians' capacity to conceptualise, we shall see here whether the latter two components may be disentangled sufficiently to allow machines to augment our capacity to prove and conjecture. In chapter 3 I deal with automated theorem proving, contrasting two approaches: one which aims to emulate human reasoning and one which sees fit to play to the strengths of machines at matching syntactical strings. It is the second approach which has produced the most striking result to date by contributing to the solution of a problem which taxed Alfred Tarski for many years.

Continuing the theme of computerised assistance, in chapter 3 we see what has been achieved in the line of conjecture formation. One of the most effective areas is the use of high-precision arithmetic to discover numerical identities between combinations of constants. Other successes include the use of algorithms to count tilings in enumerative combinatorics. What I shall be focusing on in this chapter, however, are my own attempts to use a logic-based machine learning device to learn from a collection of topological data. I shall argue that the difficulties I encountered arose from the unsuitability of a logical language in this topological domain.

Induction and analogy were the two varieties of plausible reasoning George Pólya treated in his *Mathematics and Plausible Reasoning* (1954a). Having looked at inductive activity in chapter 3, in chapter 4 I turn my attention to analogy. We can find ample evidence of the high regard in which analogical reasoning is held by mathematicians, but do its apparently successful uses indicate merely that it represents a way mathematicians can think powerfully, or rather that it is the style of thinking best adapted to capturing inherent structure? Is it possible to transfer constructions between any two fields of mathematics, or are we encountering what we might call after Wigner the *unreasonable effectiveness of mathematics in mathematics*?

Part II considers plausibility, uncertainty and probability. In chapter 5 I probe Pólya's idea that, like everyday reasoning, mathematical reasoning is for the larger part inductive rather than deductive. In explaining this idea he chose to represent uncertainty in mathematical knowledge using a kind of Bayesian degree-of-belief framework, where qualitative changes of direction of degrees of belief occur on discovering new knowledge. Now, while Bayesian reconstructions of rational practice are extremely common in philosophy of science, very little has been done by way of following up Pólya on their suitability for mathematical practice. I discuss the question of which variety of Bayesianism is most appropriate for this task, and proceed to examine the use of analogy, the choice of strategy in proof planning, and large-scale enumerative induction.

If mathematical uncertainty may be viewed in Bayesian terms just as scientific uncertainty often is, we should expect a certain compatibility between the way we treat them. Today, in the absence of experimental data, it may happen that the only verifiable predictions are mathematical ones. When these mathematical predictions are confirmed, scientists naturally view this as a sign that their modelling techniques and theorisation are on track. I argue in chapter 6 that any confirmation theorist must either declare this irrational or else integrate mathematical predictions into their framework. For the Bayesian philosophy of science, the obvious solution is to extend Bayesianism to mathematics as I did in chapter 5.

In part III, Chapters 7–9 concern Lakatos: what he did, what he might have done but for his untimely death, and what he perhaps ought to have done. In chapter 7, I examine his writings in philosophy of mathematics. Lakatos maintained that the transformation of informal mathematics into an axiomatised research programme largely marks the end of the creative process. While indicating that readjustments might be required were insufficient of the informal material to be captured, he gave no hint that an axiomatisation could act as a springboard for further theoretical development. For Lakatos, once a mathematical theory has entered the axiomatic stage 'imagination is tied down to a poor recursive set of axioms and some scanty rules' (1978b: 68). I argue that Lakatos is wrong.

It is known that Lakatos had intended to return to the philosophy of mathematics to apply the constructions he had devised to account for the development of the natural sciences. In chapter 8 I assess the obstacles to a transfer of the methodology of scientific research programmes to mathematics. I argue that, if we are to use something akin to Lakatos's methodology to discuss modern mathematics with its interweaving theoretical development, we shall require a more intricate construction and we shall have to move still further away from seeing mathematical knowledge as a collection of statements. I also examine the notion of rivalry within mathematics and claim that this appears to be significant only at a high level.

Out of the totality of possible mathematical structures, mathematicians choose to explore a very limited range. This is due not solely to what they are able to study at any one time, but rather it reflects their attribution of importance. Naturalness, fruitfulness, conceptual power and elegance are other qualities valued in a piece of mathematics. Are these merely transitory aesthetic judgements, or is there something 'objective' to them? In chapter 9, I consider the claims made by some mathematicians that conceptual development is afforded too little credit by comparison to problem solving capacity.

In part IV, in the final chapter (chapter 10), I take a look at higher-dimensional category theory. This new discipline seeks to construct higher-dimension versions of simple algebraic constructions by a process known as *categorification*. An example of this proceeds up the ladder of categorification from the integers to the category of tangles in which sits the collection of mathematical knots. In the process it explains why mathematicians and physicists have found knot invariants by way of a construction on types of Hopf algebra, known as quantum groups, and relates this construction to models employed in statistical mechanics and quantum field theory. One intriguing consequence of its adoption is a reappraisal of sets as occupying simply one corner of the mathematical universe.

As you will have gathered, the chapters of this book cover a considerable range of aspects of mathematics. This is as it must be. If physics can meet philosophy at the Planck scale, in time I hope that philosophy and mathematics will meet in a very wide range of places. We should broaden the range of philosophical resources in these encounters. The ones I have used originate principally in the English-language literature, and at the 'internalist' end at that. This should not be taken to imply a want of appreciation for what is happening elsewhere, but when forging a path towards a philosophy of real mathematics one must start somewhere.

PART I

Human and artificial mathematicians

Future philosophers may find it surprising that so few analytic philosophers, especially those who work at the formal end of the spectrum, have troubled themselves to converse with practitioners of machine learning, seeing that both groups convey their ideas about knowledge and inference within logical systems. Mathematical reasoning has, of course, been seen as the paradigm case for this treatment, so it is interesting to read Seymour Papert, a prominent Artificial Intelligence researcher, accuse philosophers of working with:

a projection of mathematics which greatly exaggerates its logical face much as the Mercator projection of the globe exaggerates the polar regions so that on the map northern Greenland becomes much more imposing than equatorial Brazil. (Papert 1978: 107)

One reason for my interest in computer assisted mathematics is that it offers us our best chance of perceiving what is lacking in such a projection.

Theorem proving, conjecturing and concept formation make up the three principal components of mathematical research. The brilliant observation of Lakatos, argued through the length of his famous dialogue, *Proofs and Refutations*, was that these components are thoroughly interwoven. He was right. Mathematicians perform these activities simultaneously – while clarifying a concept they notice a property which looks like it may hold for all of some class of objects, and while trying to prove that this is so, they find that it pays to introduce conceptual distinctions between elements of that class.

The use of computers in mathematics up until now has been surprisingly limited, but the evidence of the past few years indicates that some changes are on their way, especially with the advent of computer algebra systems, to augment number-crunching capabilities. For computers to become valuable to mathematicians it will be necessary to isolate the components, or perhaps even parts of the components, mentioned above. Numerical and symbolic calculation may be seen as an integral part of theorem proving, conjecture formation and conjecture testing. The last of these three has been far and away the most successful avenue for computer science to invade mathematics, but we should note that any success here need not make us change our image of mathematics so very radically if we were already aware of mathematics' quasi-empirical face. We should already have known that numerical calculation may be used to put conjectures to the test. Computer algebra systems just allow us to expand this activity.

I want then to look here at potentially more radical uses of machines. Computer science provides the philosopher of mathematics with a fresh angle on mathematical practice, the chance to look outside of what might turn out to be merely 'human' mathematics to a larger space of ways of doing mathematics. In chapters 2 and 3 I shall be exploring what we might call 'fringe' uses of machines in mathematics, in theorem proving and conjecture formation, as a way of deepening our understanding of what doing mathematics at the start of the twenty-first century is like.

The final component of mathematical research activity I spoke of above is concept formation. I shall not be treating 'beyond the fringe' attempts to automate this process owing to a near total lack of success to date. After the initial flourish of Lenat's *Automated Mathematician* (AM), little has been done here. One of the few recent attempts has been the HR program, which 'genetically' recombines pieces of other concepts (Colton 1999). One of its more notable feats has been to generate the notion of a *refactorable* number. A refactorable number is a natural number the number of whose divisors divides itself. The first few are 1, 2, 8, 9, 12, 18 . . . While this series has been included in an electronic encyclopaedia of integer sequences, I doubt many mathematicians will be impressed. So, instead, in chapter 4 I see how humans go about concept formation via analogy, to provide a counterpoint to chapters 2 and 3.

CHAPTER 2

Communicating with automated theorem provers

> it is only the very unsophisticated outsider who imagines that mathe-
> maticians make discoveries by turning the handle of some miraculous
> machine.
> (Hardy, A Mathematician's Apology)

2.1 INTRODUCTION

Had you subscribed to the *New York Times* back on December 10, 1996, you might well have noticed the following headline:

Computer Math Proof Shows Reasoning Power
By Gina Kolata
Computers are whizzes when it comes to the grunt work of mathematics. But for creative and elegant solutions to hard mathematical problems, nothing has been able to beat the human mind. That is, perhaps, until now . . .

The article announced that a computer had solved a famous mathematical problem – The Robbins Problem – sixty years after it had been posed. Noted mathematicians had tried but all had failed. Even the great logician Alfred Tarski had spent time on it to no avail.

To date computers have had little impact on the process of deriving mathematical proofs, or, at least, very much slighter an impact than one might casually have reckoned on from the way mathematics was repre-sented in much of the philosophical literature of the twentieth century. To give a couple of examples briefly, where Pierre Duhem spoke of the role of *bons sens* and *finesse* in the field of physics, he contrasted these to *géométrie*, the automatic mode of thought to which the mathematician is restricted (cf. Crowe 1990). For the logical empiricists, meanwhile, mathe-matics was taken to be merely a branch of logic, give or take an odd pos-tulate (Hempel 1945, §10). Duhem's idea of automaticity was absent from Hempel's account in recognition of mankind's psychological limitations, but bearing in mind that computers excel at syntactical manipulations corresponding to steps of logical inference, which includes for Hempel *mathematical* inference, we might have expected by now the extensive use of banks of automated theorem provers in our mathematics departments. What we find instead, however, is that mathematicians are using computers for quite other purposes. Aside from their more menial functions in facilitating communication and publication, several areas of mathematics have benefited greatly from computers' number-crunching capabilities and could hardly have been developed without them. I am thinking here of, for instance, numerical analysis of partial differential equations. In a similar vein, as we shall see in chapter 3, there are signs of an emerging experimental approach where researchers can 'out-Gauss' Gauss by carrying out lengthy numerical and algebraic calculations to see whether their conjectures hold over a range of instances, or whether patterns emerge from their data.

What has not happened yet has been any appreciable turn to computers to assist in the production of the inference steps of a proof. While a degree of success has been achieved by theorem provers, prompting a few leading mathematicians to predict a rosy future for the automation of theorem proving, at present it is very much a fringe activity for the mathematical community, and we may expect that it will remain so until a significant number of problems deemed to be of sufficient importance have been resolved. Notwithstanding this lack of interest on the part of mathematicians, the ambitions of researchers to develop powerful automated provers and proof assistants remain undiminished.

In spite of its marginal status, an investigation of automated theorem proving remains worthwhile since it sheds light even on central aspects of the mathematical community's research activity. Our primary concerns are the following. First, a simple general remark, one should not underestimate the difficulty of developing useful automated theorem provers. Even in the case of first-order provers, a huge effort has been required to find strategies to restrict and direct machines to be effective even on a very limited range of problem types. Second, I believe it is no accident that the most successful approach to date has been one that has deliberately avoided closely imitating human problem solving techniques. Computers have their own inhuman strengths which need to be harnessed. Third, the output of a theorem prover, even when it represents the proof of an open problem, is often of little value beyond establishing the correctness of a proof. Human mathematicians pride themselves on producing beautiful, clear, explanatory proofs, and devote much of their effort to reworking results in conceptually illuminating ways. Philosophers must not evade their duty to treat these value judgements in mathematics. Fourth, it is worth considering that automated provers might allow mathematicians to develop mathematical domains which they would not otherwise have done. Perhaps, not only have mathematicians been constrained to study subject areas in which they could calculate, expansion occurring through theoretical, algorithmic and technological advances, but also they have been limited to domains in which results are humanly provable. Finally, if anything very important is to come out of this research, mathematicians will have to learn and devise new language forms, both to be able to speak to members of the automated reasoning community and their machines, and to glean as much as possible from the latter's often obscure output.

In the next section I shall briefly outline ongoing work in the automated reasoning community. In section 2.3 I shall discuss the approach to theorem proving adopted by Larry Wos's group at Argonne, Illinois, at present the most successful group working on mathematics and logic. Their finest achievement to date has been to settle the Robbins Problem, sixty years after it was formulated. This is the topic for section 2.4. Their interest piqued by a proof which seems to lie just a little beyond the grasp of humans, various researchers have attempted to reformulate it in a more humanly accessible way. Finally, in section 2.5 I discuss the knot theorist Louis Kauffman's use of a diagrammatic notation for this purpose.

2.2 AUTOMATED MATHEMATICAL REASONING

Leibniz's dream that some form of unambiguous symbolic language could be devised to allow all forms of reasoning to become mere calculation is alive and well among practitioners of the field of automated reasoning. To date the most solid achievements in this field have been the construction of systems that can automatically synthesise hardware and software designs to meet given specifications, or that can verify that such specifications have been met. Here, electronic circuits and computer programmes are represented as logical formulae, allowing their features to be interrogated.[1] However, the automated reasoning community has more ambitious goals. For example, a considerable number of researchers are interested in applying

---

[1] See MacKenzie (2001) for this and other issues bearing on the chapter.

automated deduction techniques to natural language processing tasks. By contrast to the widely used statistical approach, from this perspective the act of parsing a sentence is seen as a form of logical deduction, where choices must be made among the various pieces of typing information tagged to the different senses of each word in the sentence in order to produce a consistent assignment. When a parser reaches the end of a sentence still expecting a noun phrase or anticipates a full stop too soon, it 'realises' that it has made this choice incorrectly and must backtrack. An idea of this can be gained when, for instance, English speakers typically have to adjust their parsing in the case of the sentence – *The horse raced past the barn was fast* – by redesignating 'raced' from an active intransitive verb to a passive transitive one heading a subordinate clause.

Our concerns here, however, are with one of the original aims of automated reasoning – mathematical theorem proving. Ambitious plans are afoot, and have partially been achieved, to link together several theorem provers, model constructors and databases of known theorems to form a mathematical assistant. Mathematicians would be able to pose problems from their desks to an interface program via the Internet. The central node of the network distributes the request to several available client theorem provers, while the interface shows the user the state of progress. This network would assist in devising a proof plan, which would isolate potentially important lemmas, and attempt to prove them, drawing on knowledge contained in databases.

Before we get carried away by these enticing visions, it is worth looking to see what has been achieved by automated theorem provers to date. We have had to learn to be very cautious about the exaggerated claims of practitioners of artificial intelligence in the past, their failure to deliver justifying a considerable degree of scepticism. An insight into their achievements may be acquired from a perusal of the database of problems against which automated theorem prover designers test their programs. A recent version (v.2.3.0) of the TPTP (Thousands of Problems for Theorem Provers) repository contains around 3,500 files, each containing information about a problem, although some concern the same problem and many are minor variations of others.[2] One of the largest groups of problem concerns group theory (see appendix), interpreted broadly to include topics such as quasi-groups, typical members of which are:

(a) The inverse of the identity is the identity.
(b) The left identity is a right identity.

---

[2] This may be consulted at http://www.cs.jcu.edu.au/~tptp/.

(c) The identity element is unique.

(d) If $a = b^{-1}$, then $b = a^{-1}$.

(e) $(x \cdot y) \cdot (y^{-1} \cdot x^{-1}) = e$.

(f) If G1 has exactly two elements and G2 has exactly two elements, then there exists an isomorphism (a one-to-one and onto homomorphism) between them.

(g) Groups in which the square of each element is the identity are commutative.

(h) For groups in which the cube of each element is the identity, then for all x and y, we have $[[x, y], y] = e$, where $[a, b] = aba^{-1}b^{-1}$ is the commutator of a and b.

(i) The commutator operation is associative if and only if the commutator of any two elements lies in the centre of the group, i.e., $[[x, y], z] = [x, [y, z]]$ iff $[u, v] \cdot w = w \cdot [u, v]$.

(j) There is a single axiom for group theory, in terms of product and inverse.

The first five, (a)–(e), are elementary results, each of which has a fairly clear meaning that could be used in an informal proof. For example, as when taking off socks and shoes, the idea behind (e) is that to undo the composite action, A followed by B, one must first undo B then undo A. However, even an elementary textbook will give fairly formal derivations in the form of the rewriting of equations. Similarly, a textbook presentation of the final four statements, which from the simple (g) to the tricky (j) and forward half of (i) lack intuitive obviousness, would adopt this equational form. All of these results are thus well suited to theorem provers working with equational first-order logic.

(f) presents an interesting case in that it is the type of problem that is a very obvious result for a human and yet it is difficult for a machine. Unlike the other cases where human proofs will resemble computer proofs as the rewriting of equations, here the human will typically be rather informal. For instance, she might point out that if the groups elements are e (identity) and a, then three of the entries in the multiplication table are instantly determined by the properties of the identity, and that since a needs an inverse and $a \cdot e \neq e$, then we must have $a \cdot a = e$. A final comment to the effect that the multiplication table of any other two-element group will be forced to take the same form up to isomorphism completes the argument.

The computer, on the other hand, must reason at a more logically explicit level. The comments from two files recording computer attempts to solve it are revealing. The first of these states that:

In order to prove the theorem, the group tables and a particular homomorphism are specified, and the contradiction comes from the fact that this is the actual isomorphism. Not only is this formulation cheating, but also it does not prove the theorem in full generality. (TPTP, file GRP025-1)

Presumably, the worry is that all that has been shown is that two particular two element groups are isomorphic, the mapping having been provided. That a serious attempt to prove such a simple result should fail to establish it in full generality, and should involve so generous a hint to the machine as to be described as 'cheating' is surely revealing. A second attempt is more successful, but still requires generous help:

In order to prove the theorem, we specify one element of each group as the identity element and take as a previously-proven lemma (obvious) that maps from $G_1 \to G_2$ which are not one-to-one or which are not onto need not be considered for isomorphisms between the groups. Thus we consider only the two one-to-one and onto maps between the groups, and show that assuming neither of them are homomorphisms gives a contradiction. (TPTP, file GRP025-2)

What this shows is the difficulty of higher-order reasoning for present-day machines. Problems more naturally expressed in higher-order terms are either represented unnaturally to be fed to a first-order logic theorem prover, and in the process helped considerably, or else they are fed to a higher-order logic theorem prover which almost always needs much assistance throughout the proof. Some researchers seem to be content with this situation, aiming to design what we might term a 'secretarial' theorem prover which acts merely to fill in the short gaps of a proof plan devised by its human user; others are more ambitious.

Recognising that human mathematicians think in terms of proof plans, work has been carried out on automating their construction. However, for this to be feasible there has to be a sense of a 'typical' situation where a particular proof strategy applies. This is fair enough in the verification of chip and software designs, where one finds great similarities in the required strategies, where nothing very novel is expected and where the outcomes are long tedious proofs of specifications being met, which have no intrinsic interest to anyone but the designer. One could imagine this approach to automated theorem proving arriving eventually at a machine that would do well in a public examination where increasingly, in Britain at least, the student is expected to have learned to apply a limited range of techniques in a near algorithmic way. This would admittedly be an enormous achievement, but even then it is not clear that we would be very much nearer to a device capable of assisting working mathematicians. Time will tell. As I am interested in the production of new results I shall be devoting much of the rest of this chapter to the theorem provers produced by Larry Wos and his team, in this respect the most successful. But to throw Wos's approach into clearer relief, it will be worth our while ending this section by touching on an example of the proof planning approach.

I shall draw here on an article in the *Journal of Automated Reasoning* by Erica Melis (Melis 1998), in which the author explains her approach to automating analogical translation of a proof plan between two similar problems. The motivation for her work was a challenge made by Woody Bledsoe, a mathematician very interested in automated theorem proving, who set the automated reasoning community the task of producing a proof of the Heine–Borel theorem in two dimensions given a suitable proof of the corresponding result in one dimension. This theorem, in its standard one-dimensional form, states that for any closed interval of the reals, [*a*, *b*], any covering by open sets has a finite subcover. In the two-dimensional case one replaces the interval by a closed rectangle. Think of covering a table top with infinitely many mats, where overlapping is allowed and there is no limit as to how small a mat may be. The theorem tells you that you never need to use that many.

Bledsoe's intuition was that this transfer is an elementary, yet typical piece of mathematical activity. If for mathematicians the step from one dimension to two is straightforward, then any automated theorem prover worth its salt had better be able to do likewise. However, a direct process of syntactical translation is doomed to failure. Indeed, when you attempt such a feat, you can easily see what Taylor means when he says 'the predicate calculus has a better claim to being the "machine code" of mathematics than set theory or the Sheffer stroke does, but machine code is always rather clumsy in handling higher level idioms' (Taylor 1999: §2.6.9), emphasising its distance from the level of mathematical ideas. Instead, one must represent the proof ideas in a higher-level description language.

The strategy of an often-quoted proof of the one-dimensional Heine–Borel theorem is to seek a refutation by assuming that there is a closed interval and an open cover with no finite subcover. Divide this interval into two equal parts and select the leftmost one which cannot be finitely covered. Iterate this process to achieve an infinite sequence of nested intervals, each of which is half the length of its predecessor and cannot be finitely covered. The key now is to use knowledge of the reals to the effect that the intersection of such a sequence must contain a point. As it is contained in the initial interval, this point must belong to some member of the open cover. This open set must contain a closed interval centred on the point, but far enough along the sequence the nested intervals are narrow enough to fit inside this closed interval, implying that they have a one-element cover, which contradicts the fact that they could not be finitely covered.

Mellis represents this proof in a tree-shaped format as a *proof plan*. To a goal or target proposition there is associated a method. Each method relies in turn upon a branching set of assumptions (lemmas, axioms, definitions) and further goals. Each of these further goals is in turn to be achieved by a method with its own branching set of assumptions and new goals. For a valid proof plan the formulas represented at the leaves must all be assumptions.

What Mellis is trying to achieve is the automated translation of the proof plan to the two-dimensional case. Her program thus rewrites, say, mention of 'closed intervals of the reals' in a goal or assumption to 'closed rectangles of the plane'. The new versions of the lemmas must then be checked, along with the translated proof methods. In this case, the part that needs most adjusting occurs when showing that eventually members of the sequence of nested rectangles are contained in a rectangle centred on the point contained in the intersection of the sequence. Here one must ensure that one is far enough along the sequence that both length and breadth are small enough to fit in that rectangle.

This is impressive work, but still very much in the secretarial mould. The operator has to provide the theorem, proof plan and relevant definitions and lemmas of the source of the analogy, along with the theorem, definitions and lemmas of the target. You have to remind yourself that given the proof in one dimension in terms of a nested sequence of intervals, it is straightforward to use the very similar idea of a nested sequence of rectangles generated by quartering each time and choosing one of the quarters not finitely coverable. The machine is labouring to assist in a task which for the human with the slightest grasp of the situation is obvious. To be in a situation where one has a sufficiently elaborated proof plan for a result makes it unlikely that one would not be able to carry out very easily an analogical transfer oneself.

Of course, we should not be too quick to judge. Perhaps, these are just the first faltering steps of a young research programme striving to construct a useful computerised assistant. It is interesting to note, however, what Bledsoe wrote in 1986:

Analogy is the heart and soul of all intelligent behavior, especially mathematical behavior. Why have we made so little use of analogy in ATP [automated theorem proving]? . . . We predict no substantial advance until our provers begin to effectively use Analogy with the help of an adequate MKB [mathematical knowledge base]. (Quoted in Melis 1998: 255–6)

Ten years later a long-standing problem had been resolved, generating a large amount of publicity, but by an approach with no pretensions of emulating human mathematicians. As we shall see, the philosophy adopted by Wos and his team has been very much to play to the computer's strengths.

2.3 THE ARGONNE PARADIGM

The Argonne approach to automated theorem proving is characterised by its experimental outlook. The key for Wos's group has been not to imitate slavishly the human. Rather, he has sought to represent problems in such a way as to allow the computer to carry out inference in the way best suited to it, but at the same time allowing the human expert to influence the inferential process.[3] Several inferential strategies have been devised for this purpose. The user of OTTER must then choose which strategies to employ on a particular problem and which settings to apply regarding these choices. It has been made very simple to experiment by altering these choices and settings to see whether a proof can be found,

A first indication that we must play to the machine's strengths is the use of the *clause language*. The program represents sentences as disjunctions of positive and negative literals. Logical 'or' is represented by '|', while 'not' is represented by '−'. Thus, the following sentences are acceptable clauses:

```
-P(x, y, u) | -P(y, z, v) | -P(u, z, w) | P(x, v, w).
-P(x, y, u) | -P(y, z, v) | -P(x, v, w) | P(u, z, w).
```

By attempting to understand what these sentences amount to in the context of P expressing a product relation, the reader will no doubt conclude that clause language is no natural vehicle for the expression of mathematical ideas.

This situation is ameliorated by the capacity of OTTER to accept sentences in standard predicate form. The use of Skolem functions permits a translation from the first-order predicate calculus to clause language, where the resulting set of clauses is not necessarily logically equivalent to the original statement or formula, but the original statement is satisfiable iff

---

[3] This dichotomy between the imitation of human reasoning patterns and the reliance on the sheer speed of a computer to search through a massive space has been with us right from the beginning of artificial intelligence. See MacKenzie (2001, ch. 3).

the translation is. OTTER is able to perform this translation, and would translate the following sentences to those above:

```
all x y u v z (P(x, y, u) & P(y, z, v) & P(u, z, w)
  -> P(x, v, w)).
all x y u v z (P(x, y, u) & P(y, z, v) & P(x, v, w)
  -> P(u, z, w)).
```

This is surely an improvement, but by no means a perfect situation. First, any constructed proof appearing in an output file will still be represented in clausal form and so require translation into first-order form.[4] Second, and more importantly, mathematically one would prefer the use of equality based reasoning. OTTER does accept such a representation. In the case of the sentences above it takes the far more familiar form:

$$\text{EQUAL}(f(f(x, y), z), f(x, f(y, z))),$$

revealing itself to be an expression of associativity. However, adopting this familiar form forces you to use the type of inference known as *paramodulation*, rather than any variant on the classic *resolution*.

Perhaps the clearest way to gain a sense of how what is best fitted to the computer may be very different from what is suitable to the human mathematician is to consider a particular reasoning manoeuvre. Were a mathematician asked to show that any group all of whose elements square to the identity is commutative, it would probably not be long before they thought to substitute *yz* for *x* in $f(x, x) = e$. However,

no effective automated technique is known for wisely choosing which of the myriad of less general conclusions to draw, indeed, how to effectively emulate that aspect of person-oriented reasoning. In other words, automated reasoning programs do not offer the type of reasoning called *instantiation*, which can be used to yield the second equality from the first by replacing (instantiating) *x* by *yz*. Although instantiation serves well logicians and mathematicians, unless an effective strategy is discovered to control its use, instantiation is unneeded and even unwanted in the context of mechanizing inference rule application and proof finding. Indeed, its use (in effect) conflicts with a reasoning program's preference for generality that in turn contributes to effectiveness. (Wos and Fitelson forthcoming: 5)

For a device ever threatened to be swamped by derived consequences, the last thing it needs is a capacity to instantiate willy-nilly. Instead, it needs to find a way to combine two or more clauses it already has in the hope

---

[4] The LΩUI (Lovely Omega User Interface) is capable of automatically translating a proof from clausal form to something more comprehensible.

of coming up with something new. If, for example, OTTER has already generated:

(1) $\underline{f(f(x, y), y)} = x$,

(2) $f(x, \underline{f(x, y)}) = y$,

then it *can* choose to assign x in (2) as $f(x, y)$, since this permits unification of the underlined terms, yielding

(3) $f(f(x, y), x) = y$.

This step is an instance of what Wos calls paramodulation:

> Of the various inference rules, one of the more complicated is *paramodulation*, which enables an automated reasoning program to treat equality as if it is 'understood'. Paramodulation – which is the best example of a computer-oriented inference rule, and one that a person probably should not apply by hand – generalizes the usual notion of equality substitution. (Wos and Fitelson forthcoming: 9)

So the generation of new clauses is prevented from overloading the system by the control of instantiation, using techniques such as paramodulation, and by the deletion of less general clauses, using the closely related demodulation.

At the same time as it is possible to *restrict* OTTER's inference, it is also possible to *direct* it. An example here is given by the resonance strategy. Unless otherwise directed, OTTER weights clauses according to the number of symbols appearing in them. The resonance strategy allows one to intervene by telling OTTER to give a lower weighting to clauses possessing specified syntactical forms. Thus, if you believe that a proof is possible via a certain lemma, you can ensure that OTTER looks favourably on any clause of the same syntactical form. It is interesting to note in this context that Wos uses the phrase *resonance strategy* for what he once termed *analogy*. If you have a proof for a proposition and you come to consider a very similar proposition, you can direct OTTER to give low weights to generated clauses fitting syntactical templates corresponding to lines of the original proof. Notice how this technique contrasts markedly with Melis's analogical reformulation that we saw earlier.

With resonance and similar strategies available, the user has many choices to make to steer the machine to succeed either in producing a first proof of a result or in producing a shorter proof than any previously known. From my amateur attempts at automated proof, I found it very striking how sensitive OTTER is to these choices. For example, you may think you are helping it on its way but find you have side-tracked it into paying too much attention to clauses of a specified syntactical form. Then again, an adjustment may have the effect of increasing the CPU time considerably, but also of producing a shorter proof. OTTER operators clearly develop a practical knowledge of what is likely to work, which exceeds what they are able to convey by rules of thumb. I find it unlikely that difficult problems will be solved by OTTER without considerable input from someone who has worked closely with Wos's team. This was certainly true of its most fêted result – the Robbins Problem.

2.4 THE ROBBINS PROBLEM

In 1933, Huntingdon presented a simple set of axioms for Boolean algebra. He demonstrated that in addition to specifying that there be a commutative, associative binary operator, denoted '+', and a unary operator, denoted 'n', all that was needed was an axiom relating them in an equation:

(1) $x + y = y + x,$ (commutativity)

(2) $(x + y) + z = x + (y + z),$ (associativity)

(3) $n(n(x) + y) + n(n(x) + n(y)) = x.$ (Huntington equation)

From these it is possible to show the existence of a '0' for the addition, that $n(n(x)) = x$, that $n(x + n(x)) = 0$, the existence of a '$\times$' operator, distributivity laws, etc. Shortly after Huntingdon's paper appeared, Robbins speculated that the Huntingdon equation could be replaced by a weaker one:

(4) $n(n(x + y) + n(x + n(y))) = x.$ (Robbins equation)

It is easy to show that a Boolean algebra satisfies the Robbins equation, but it remained an open problem for around sixty years to determine whether a Robbins algebra, i.e., one satisfying (1), (2) and (4), is Boolean, and this despite the best efforts of Tarski and his students.

Wos and his colleagues had been aware that the Robbins Problem was one well suited to their techniques several years prior to the proof. A large part in its solution was played by a mathematician named Steve Winker, who at Wos's suggestion sought conditions which if satisfied by a Robbins algebra would imply that it was Boolean. The idea was to use a simple condition of this kind as a lemma and then see whether this lemma was derivable solely from the Robbins axiom. Several candidates were found, but most seemed to offer little prospect of having made the problem simpler in that they looked just as hard to prove as the Huntingdon equation did itself. However, a few promising ones did eventually emerge around 1980. The two simplest conditions Winker found were:

(i) there are two terms, C and D, satisfying the relationship $C + D = C$,
(ii) there are two terms, C and D, satisfying the relationship $n(C + D) = n(C)$.

Winker had proved these conditions sufficient using induction, something unavailable to OTTER, but by 1996 Wos's group had used OTTER to generate first-order proofs that Winker's conditions sufficed. William McCune then took up the task of searching for two terms satisfying this property. For the purpose he used the automated reasoning program EQP, a variant of OTTER which could perform AC-unification. Here instead of writing in the commutativity and associativity of the '+' by hand as axioms, the program automatically looked to unify and subsume terms by commuting and reassociating subterms. On one run of EQP lasting eight days in October 1996, two terms were found satisfying the second condition. Shortly after, EQP found two terms satisfying the first condition as well as a resolution of the Robbins Problem which avoided Winker's conditions.

To give a taste of one of the computer proofs, we shall look at its first step. Recall that the Robbins equation takes the form:

$$n(n(n(w) + z) + n(w + z)) = z.$$

A use of paramodulation involves replacing w by $(n(x) + y)$ and z by $n(x + y)$ to yield:

$$n(n(n(n(x) + y) + n(x + y)) + n(n(x) + y + n(x + y))) = n(x + y),$$

which reduces to

$$n(y + n(n(x) + y + n(x + y))) = n(x + y),$$

on rewriting the first term within the first pair of parentheses on the left-hand side according to the Robbins equation. The proof continues in this vein until after only twelve steps it is shown that

$$n(n(3x) + x) + 2x = 2x,$$

where mx is the sum of m copies of x. Winker's first condition has been satisfied.

In some sense, it seems fitting that the subject matter is Boolean algebra. After all, one of the earliest examples of an automated reasoner was Jevons' *reasoning piano* which could determine the truth status of Boolean algebra expressions. But Boolean algebra was devised in the nineteenth century with an intended interpretation, namely, propositional logic, and was later found to be very useful for understanding circuit design. Why then would one want to solve the Robbins Problem? What is to be gained from seeing how weak an axiom base one can get away with? Much is made of the fact that Tarski directed a considerable amount of attention, both his own and his students', to the problem, but this fact cannot be decisive. A considerable portion of the automated theorem provers' successes have been in establishing that a smaller number of axioms can form the basis of an algebra than had been thought. It is true that mathematicians have engaged in these games, devising single axioms for various algebraic objects. Tarski in 1938 showed that a set with a binary operation '−' satisfying

$$x - (y - (z - (x - y))) = z,$$

is an Abelian group, while Higman and Neumann showed in 1952 that a set with a binary operation '/' satisfying

$$(x/((((x/x)/y)/z)/(((x/x)/x)/z))) = y$$

is a group. One may worry, however, that this is not a very worthwhile pursuit. It might have been the case that the fewer the axioms describing a field, the easier it would be to establish theorems there, in the sense that one might expect less combinatorial explosion. But of course the proof that EQP generated establishes just the opposite. It takes twelve intricate lines of proof to establish that there exists a pair of elements whose sum is equal to the first of them, something immediately deducible from the usual axiom basis. Certainly, Wos sees no point in handicapping a system by restricting its axiom base:

My experiments suggest that, as is so often true in mathematics, using a minimal set of axioms can actually interfere with the likelihood of success. For some evidence, you need merely glance at a typical algebra text in which group theory is discussed. Specifically, although the axioms of right identity and right inverse can be proved dependent on the remaining set consisting of left identity, left inverse, and associativity, the typical author simply includes the dependent axioms when discussing group theory. (Wos and Pieper 1999: 376)

On the other hand, it is possible that we might wish to check that a particular set with a binary operation is an Abelian group. Because of Tarski's result we would only have to check that his single condition holds. However, I have never seen this done.

Perhaps we can say nothing more than that EQP has given us a new insight into Boolean algebra. The experience gained from working on the Robbins Problem may help in less-charted areas of mathematics. We can anticipate that mathematicians and scientists arriving at a situation where they would like to know more about a particular algebra but have little intuition about it will turn to automated provers such as OTTER and EQP for assistance. It will then become important to learn not only the correctness of conjectures about the algebra, but also how to achieve further such results by becoming acquainted with how the elements of the algebra behave. To facilitate this process will probably require translation procedures to more humanly comprehensible forms. Let us now turn to consider attempts to do this for EQP's proof.

2.5 REFORMULATION

If the day arrives when automated theorem provers become powerful enough to be able to resolve open problems across a significant spectrum of mathematical disciplines, how much shall we have benefited? Knowledge of the truth or falsity of a proposition of interest is, of course, useful, but, as Rav (1999) has discussed, it is far from everything.[5] Mathematicians want to learn much more from each other's proofs than the correctness of their conclusion. Indeed, a good human proof will typically introduce important lemmas and novel concepts which may turn out to be more significant than the theorem itself. For instance, a good 'conceptual' proof of the irrationality of $\sqrt{2}$ will make you see that for n a positive integer, $\sqrt{n}$ is rational if and only if every prime factor in the unique prime factorisation of n has an even exponent, i.e., n is a perfect square. Similarly, an aptly chosen counter-example is worth probing for further information.

Useful concept production appears to be a distant prospect, but we would be content with lesser successes. Certainly, when, as has happened, a computer's proof or disproof of a proposition prompts a mathematician to

---

[5] Rav overstates the case by questioning the value of an oracle which could answer any mathematical problem posed to it. With such a thing available one could pose candidate lemmas to it. Knowledge of which are correct would surely help in planning a strategy for a proof.

produce their own proof or counter-example just because she knows which is the correct thing to do, then we can say that the machine has played an important role. But what would be more helpful would be the chance to augment our understanding of a domain by examining the computer's output. When a computer algebra system performs a calculation beyond human capacity, we may be very interested in the result, but we are not interested in the path of the calculation. No selection has occurred in the application of its algorithms. On the other hand, EQP's proof does involve the selection of a combination of inference steps from an enormous space of such combinations. Without needing to credit the machine with intelligence, there is still something to learn from devices which manage to find one or more successful paths from a very large space of possibilities. As one mathematician who has taken an interest in the Robbins Problem comments:

> EQP's proof is much more than a calculation. The proof depends upon a successful search among a realm of possibilities and the skillful application of pattern recognition and the application of axioms. This is very close to the work of a human mathematician. EQP, being able to handle many possibilities and great depths of parantheses has an advantage over her human colleagues. I understood EQP's proof with an enjoyment that was very much the same as the enjoyment that I get from a proof produced by a human being. (Kauffman 2000: 4)

The sentiment expressed here is that there is something novel about EQP's proof, something from which one might perhaps learn, rather as chess players may learn from computers' solutions to endgame positions.

Various attempts have been made to reformulate EQP's seemingly in-comprehensible proof, including Fitelson's (1998) use of *Mathematica*. But to my mind the most interesting reformulation has been devised by the knot theorist Louis Kauffman. Kauffman has been a key figure in the drive to unravel the interconnections between braid theory, three manifold in-variants, statistical mechanics and quantum field theory, the subject matter of chapter 10. His work is characterised by a very creative playfulness, which throws into question the very notion of notation. From time to time philosophers have realised that there is more to a mathematical diagram than just a heuristical aid, that in certain cases some form of access to truth is possible. As we shall see, Kauffman is part of a more ambitious movement that hopes to see the boundary between notation and diagram redrawn, even removed. Here, one can calculate and prove with diagrammatic no-tation. Representations of topological objects become pieces of algebraic notation and vice versa, each able to undergo surgical transformations to effect calculation.

This movement has amusingly termed itself *postmodern algebra*, playing on the jarring juxtaposition of the words, while alluding to the *modern algebra* devised by Noether's Göttingen school of the 1920s. Scholars of postmodernism soon discover that the attitudes of their heroes of the middle to late twentieth century are often foreshadowed by writers and artists from earlier times. In the field of literature, for example, the text *Tristram Shandy* by Lawrence Sterne is often cited as being postmodernist *avant la lettre*. If the postmodern algebraicists engage in a similar pursuit for predecessors, one person they should encounter is Charles Sanders Peirce, the American philosopher and logician. Peirce it was who devised a system for representing both (what we now term) the propositional and predicate calculus in a form which he called *existential graphs*.

In Peirce's system, an unenclosed letter, or *graph-instance*, printed on a sheet of paper corresponds to the assertion of the proposition it represents. Enclosing it in a box corresponds to asserting its negation. The juxtaposition of graphs corresponds to the conjunction of the propositions they represent. Rules of inference then permit you to perform certain specified kinds of erasure and duplication of portions of the graph. For instance, any graph-instance may be iterated (i.e. duplicated) in the same area or in any area enclosed within that. This system has been shown to be sound and complete, as has Peirce's version of first-order logic, which introduces lines, or *ligatures*, allowing for a variable free presentation.

More recently, in the 1960s a mathematician named George Spencer-Brown, in a work entitled *Laws of Form* (Spencer-Brown 1969), echoing the title of Boole's famous work, portrayed Boolean algebra and thus propositional logic in a similar form. Where many see Spencer-Brown's book as full of empty verbiage, and at best derivative of Peirce's work, Kauffman enjoys his phenomenological turn of phrase and it is Spencer-Brown's choice of juxtaposition to represent *or* rather than Peirce's *and* which Kauffman has opted for. One of the advantages of this choice is a simpler form of implication. For Spencer-Brown A → B appears as[6]

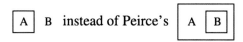

---

[6] In fact, Spencer-Brown did not use a full box notation, but just the top and right-hand sides.

Using Spencer-Brown's notation, the Robbins axiom becomes:

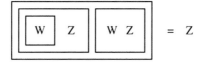

$$= Z$$

Now, remember that along with this axiom we are allowed to assume associativity and commutativity and nothing else. These, however, are built into the notation in that we are allowed to use the two dimensions of the sheet to rearrange letters, so long as we do not make them cross through the boundaries of boxes. Such an inbuilt facility seems highly desirable, as is indicated on the computer's side by the efficiency savings gained by EQP's use of AC-unification.

What paramodulation amounts to here is choosing substitutions for W and Z which make one of the inner boxes resemble the left-hand side of the above axiom.

Let us see what happens when we replace Z by $\boxed{X\ Y}$ and W by $\boxed{X}\,Y$:

The first of the enclosed boxes on the left-hand side has become an instance of the left-hand side of the Robbins axiom. We may, therefore, replace it by Y, yielding

You could imagine playing the steps of this reformulation of the proof as a film, where the passage between frames marks an inference step. As we shall see in chapter 10, postmodern algebraicists think in precisely these terms when they calculate.

Different languages suit different people. From what they have seen, some readers may be of the opinion that little has been gained by the translation into box notation. They should, though, before arriving at this judgement, thoroughly test its potential benefits for themselves, especially on some of the more intricate lines of the proof, such as

$$n(n(n(n(n(n(n(x) + y) + x + 2y) + n(n(x) + y)$$
$$+ n(y + z) + z) + z + u) + n(n(y + z) + u)) = u.$$

Kauffman clearly feels he has gained from the effort in that he is better able to appreciate the proof:

It is our contention that mathematics can behave non-trivially under change of notation. Change of notation is change of language. In the present case the change of language afforded by an appropriate change of notation makes a mathematical domain accessible to human beings that has heretofore been only accessible to computers. (Kauffman 2000: 2)

So perhaps some of us are capable of gaining a sense of EQP's proof via some diagrammatic representation. It is not clear, however, that thoroughly familiarising ourselves with Kauffman's reconstrual of OTTER's proof will help us do anything particularly useful. We should remember, however, that knowledge skills developed while working in one field may benefit work in other fields in quite subtle ways.[7] Moreover, we must look to the future:

There will be successors to EQP. Computers will prove theorems with proofs that are very long and complex. These proofs will need translations into language that human beings can understand. Here is the beginning of a new field of mathematics in the interface between logic, communication and the power to reason and understand. The machines will help us move forward into new and powerful ways to explore mathematical terrain. (Kauffman 2000: 4)

2.6 CONCLUSION

These are, of course, very early days for computer assisted mathematics. At present, by far the largest part of this assistance comes in the form of enhanced numerical and symbolic calculating capacity. In this chapter I have discussed what the marginal status of automated provers can tell us about the practice of mathematics. Even taking into account the weakness of these provers beyond a limited range of logic, the desire on the part of mathematicians for an improved understanding of a field of research discourages interest in devices generating lines of incomprehensible symbols.

[7] See Gowers (2000a).

This indifference throws into sharp contrast the dual roles of mathematical proof: establishing the truth or correctness of propositions and contributing to the conceptual development of a field. What mathematicians are largely looking for from each other's proofs are new concepts, techniques and interpretations. Computer proofs certainly give information concerning the truth of a result, but very little beyond this. However, in the combinatorial trace of their syntactical output there may be a glimmer of additional information about the theory they are working on.

This raises the further issue, the thought that there is an additional impediment to the rise of automated theorem provers, namely, that work has only just begun to find languages capable of representing mathematics to both man and machine. In his *Image and Logic* (Galison 1997), Peter Galison talks of the creation of Pidgins to facilitate communication between different communities of researchers. Just as trading partners, each with their own interests, were induced to manufacture common languages adequate for exchange, so, Galison claims, experimenters, instrument designers and theorists have found ways to communicate without the need fully to understand each other's ways. The beginnings of something similar appear to be occurring here. Kauffman is encouraging us to encode our concepts in a form acceptable to computers, and then to learn to translate from their languages to ones accessible to us. Although Galison appears to include computers within the scope of trading partners with his talk of 'Fortran Creoles', the objection may be raised that inanimate machines play no active part in language formation. Perhaps siding with Collins and Kusch (1998) in the debate over the attribution of agency in human–machine interactions, we would do better to view computer scientists (and logicians) as the mathematicians' prospective trading partners. Let us now see what assistance computers are providing for conjecture formation.

CHAPTER 3

Automated conjecture formation

Induction, i.e. inference based on many observations, is a myth. It is neither a psychological fact, nor a fact of ordinary life, nor one of scientific procedure.

(Popper 1963: 53)

The same point has sometimes been formulated by saying that it is not possible to construct an inductive machine. The latter is presumably meant as a mechanical contrivance which, when fed an observation report, would furnish a suitable hypothesis, just as a computing machine when supplied with two factors furnishes their product. I am completely in agreement that an inductive machine of *this* kind is not possible.

(Carnap 1950: 193)

3.1 INTRODUCTION

Conjecture formation stands alongside theorem proving as a fundamental kind of mathematical activity. Any prospective completely automated mathematical reasoner must not only be able to demonstrate results deductively, but must also devise worthwhile ones to feed into its prover module. The artificial intelligence community is naturally well aware of this need. Indeed, when Larry Wos (1988) raised what he considered to be thirty-three pivotal questions for workers in the field, the thirty-first of them asked:

What properties can be identified to permit an automated reasoning program to find new and interesting theorems, as opposed to proving conjectured theorems? (Wos 1988: 63)

One form the input data to such a program might take is a list of existing theorems and conjectures. Adopting a 'genetic' approach, one could perhaps feed those deemed interesting by experts into a machine programmed to mutate them slightly or to splice them together in pairs, in the hope that what was interesting about the original statements might be preserved by the offspring. Promising examples of these new statements could then be discovered by filtering out those which fail to survive a large number of tests. One might imagine, for instance, a computer being 'told' that Pythagorean triples exist, generalising to the question as to whether or not this is also the case for integral powers greater than 2. It could then easily run through millions of quadruples of integers to see whether an example of $x^n + y^n = z^n$ could be found for $n > 2$, and arrive at the statement of Fermat's Last Theorem.

As a thought experiment this may appear to be not completely implausible, but let us recognise that it is one whose relevance is heavily reliant on hindsight. That Fermat's conjecture has been such a source of fascination through the centuries has other causes than its being a variation on the theme of Pythagorean triples. From the elliptic comments of Fermat in that famous margin to the life-long quest of Wiles, much of the romance surrounding the problem stems from the way mathematicians working on the problem are publicly presented. Indeed, it should be pointed out that within the profession mathematicians have often considered the conjecture in a less than favourable light. But, however frivolous we take Fermat's conjecture to be, it has at least been possible to wed it to some real mathematics, for example, in the nineteenth century to the problem of factorisation within algebraic number rings containing roots of unity. This conglomeration of sentiments has persisted to the present day in the shape of a consensus that it is not the achievement by Wiles of his result *per se* which is important, but rather the theoretical understanding gained in the construction of the tools needed in the proof of a part, and recently all, of the Taniyama–Shimura–Weil conjecture. Without notoriety or connections to serious mathematics, conjectures will simply be ignored. Just imagine that our computer is spewing out a list of well-confirmed conjectures. It tells us, for instance, that as far as it knows an nth power added to an $(n + 1)^{\text{th}}$ power is never equal to an $(n + 3)^{\text{th}}$ power for $n \geq 17$. I predict that no mathematician would give the conjecture a second glance.

I have yet to hear of this genetic approach being put into practice, and in view of my doubts expressed in the introduction to part I about the similar HR program for concept formation, you may imagine that I am not sanguine about its chances. Let us note its Popperian flavour – conjectures arising any old how from theoretical, or even metaphysical, considerations, mercilessly tested against empirical data. Karl Popper's *falsificationism* dictates that it does not much matter how we arrive at conjectures, just so long as they are bold and readily falsifiable. Good conjectures are those which survive severe testing. What, though, of the alternative strategy of conjecture formation, namely, generalising from the data? Well, as the first epigraph of this chapter indicates, Popper was noted for his trenchant criticisms of the claim that scientific reasoning is inductive, that scientists reason from many observations to general laws.

Many commentators have been critical of Popper for conflating two types of reasoning, which might be termed inductive: finding generalisations from low-level observations and providing confirmation for general statements by verifying instances. We might call these heuristic and confirmatory induction, respectively. Already by the middle of the nineteenth century J. S. Mill in his *A System of Logic* (1843: 284) had distinguished between them. Popper's rhetorical strategy, his critics point out, was to caricature inductivists by portraying them as believing that one could proceed simply by writing down whatever one observes and then forming generalisations by an almost automatic process. Thus we read in a contemporary encyclopaedia the following entry for 'inductivism' written by the philosopher Brian Skyrms:

Inductivism, a philosophy of science invented by Popper and P. K. Feyerabend as a foil for their own views. According to inductivism, a unique a priori inductive logic enables one to construct an algorithm that will compute from any input of data the best scientific theory accounting for that data. (Audi 1995: 426)

While one can find passages of Bacon's writings which do resemble this caricature,[1] twentieth-century inductivists have been much more likely to worry about confirmatory induction. Having carved apart the context of discovery and the context of justification, logical empiricists typically kept separate the process whereby one arrived at a law or generalisation from the act of using data to provide confirmation for such a law. Concerning induction of the latter kind, Popper seemed to be on weaker ground and his ideas on corroboration and verisimilitude, the positive counterparts of falsification, have appeared to many to offer no advantage over a confirmatory inductive calculus such as Bayesianism.

For those of us living in a post-Kuhnian world where the sharpness of the context of discovery/context of justification dichotomy has been put into question, a novel and more interesting way of taking on Popper is to tackle him on his own terms. In other words, why not see whether it really might be possible to generate scientific theories mechanically from fairly raw data? Now, this is precisely what Gillies (1996) wants to find out. He takes on the controversy between Bacon and Popper precisely at the level of algorithmic methods of generalisation. He first notes that pharmaceutical

---

[1] For instance: 'There remains but one course for the recovery of a sound and healthy condition, – namely, that the entire work of the understanding be commenced afresh, and the mind itself be from the very outset not left to take its own course, but guided at every step: and the business be done as if by machinery' (Bacon 1620: 256)

companies have been engaged for decades in what he terms *mechanical falsification*. This takes place when a very large number of compounds are synthesised and then assayed for various properties with the understanding that the vast majority of hypotheses will be refuted. Like my imagined form of mathematical conjecture formation sketched above it is not a completely blind search – theoretic understanding points to certain classes of compound as being more promising than others, but still it turns out often to be more cost effective to test compounds rather than to think hard first about whether it is worth testing them. This practice might be said to lend support to Popper's attitude towards induction, since the space of promising hypotheses is not constructed from raw data in *bottom-up* fashion, but rather is reached by weeding out failures from a larger space of hypotheses devised by overarching theoretical concerns.

Gillies continues, however, by noting that machine learning programs are beginning to throw up results in an activity he calls *mechanical induction*. He outlines the case of GOLEM, an Inductive Logic Programming (ILP) program, discovering a law which indicates how an amino acid whose near neighbours on a chain possess certain properties is very likely to be part of a type of protein fold. Instead of repeating this example, let us take a look at a more recent discovery made by an ILP device. Muggleton (1999) reports that in a study of 230 chemical compounds, an earlier regression analysis had found 188 of them amenable to regression, but could make little sense of the remaining 42. His latest program PROGOL was able to provide rules allowing for equally accurate classification of the 188 compounds, but in a more comprehensible way as a set of disjunctive conditions indicative of high mutagenicity. More impressively, perhaps, it derived a rule for the remaining 'regression unfriendly' compounds which was found to be far more accurate and which indicated the 'new chemical insight that the presence of a five-membered aromatic carbon ring with a nitrogen atom linked by a single bond followed by a double bond indicates mutagenicity' (Muggleton 1999: 46). As one might imagine, this procedure is not purely bottom-up. Indeed, as we shall see below, ILP programs require background knowledge to generate worthwhile hypotheses. It can only be a question of the 'ratio', if you like, of the inductive to the deductive activity whether one calls an instance of computer assisted discovery mechanical *induction* or mechanical *falsification*.

What, then, of mathematics? Should we expect it to be possible to practice something approaching mechanical induction here? Until the advent of computers, at best, a prodigious calculator would beaver away on a few cases and conjecture some property for all n. Take the case of the Mersenne primes, those of the form $2^{2^n} + 1$. There was no systematic theory available, but still Fermat conjectured that they were prime for all n on the basis of a handful of positive cases. Now, if, as Gillies argues, mechanical induction starts to occur in scientific disciplines just as computers became sufficiently powerful to cope with significant amounts of data, we might expect that mathematicians would have begun to avail themselves of these devices. This is what I want to investigate in this chapter. In section 3.2 we shall look at calculating devices which play a role in the production of generalisations. Here machines carry out computations equivalent to the enumeration of the ways of tiling instances of various types of grid. Machines are thus shown to be playing an indispensable role in the generation of conjectures, but the algorithms used are highly tailored to the special nature of the problem situation. So next we turn to programs which conjecture numerical identities by comparing a more common type of data – long decimal expansions of expressions.

In neither of these cases, I feel, could the machines be said to performing mechanical induction. In sections 3.4 and 3.5, I shall describe a more ambitious plan to produce richly expressed conjectures automatically. As Gillies (1996) suggests, to do so we would need an automated reasoning program which uses a language capable of expressing background knowledge and data. In the light of his endorsement of the ILP paradigm, and the use of the GOLEM program to discover a law of protein folding, I have tried myself to apply the later ILP program, PROGOL, to mathematical data. Finally, in section 3.6 I shall diagnose the cause of the difficulties I had with this work.

3.2 COMBINATORIAL ENUMERATION

A graph is defined as a set of vertices or nodes along with a set of edges, each edge joining two vertices. A perfect matching of a graph G is a choice of a subset of the edges of G so that each vertex of G belongs to precisely one of the chosen edges. For example, if G is the graph whose vertices are the centres of the squares of a chess board, with edges corresponding to the adjacency of squares, a perfect matching of G corresponds to a way of tiling the chess board with blank dominoes covering two squares at a time. The question then for enumerative combinatorics is how many perfect matchings G has. A researcher in the field tells us:

For general graphs G, it is computationally hard to obtain the answer . . . , and even when we have the answer, it is not clear that we are any the wiser for knowing this number. However, for many infinite families of special graphs the number of perfect matchings is given by compelling simple formulas. (Propp 1999: 255)

Propp goes on to tell us that the desire to count matchings on graphs first arose in the 1930s for two groups of researchers. Independently, chemists studying aromatic hydrocarbons and physicists working on a theory of the liquid state were led to these enumerative problems. These interests have continued and new ones have since appeared. The field of graph matchings is now intimately related to the field of exactly solved models in statistical mechanics.

The method of attack is to translate the graph into a matrix upon which some algebraic operation is then performed. In many of the tractable cases it is the determinant of the matrix which corresponds to the number of matchings. Let us illustrate this methodology with an example. In line with having its roots in quantum chemistry, hexagonal grids, relating to the structure of benzene rings, are often studied in enumerative combinatorics. Imagine a hexagon with sides of length n, $n + 1$, n, $n + 1$, n, $n + 1$ set at $120°$ to one another, triangulated by unit equilateral triangles and with the central triangle removed. The number of ways this grid can be covered by rhombus-shaped lozenges (each covering two triangles) may be calculated by a computer program evaluating the determinant of a certain matrix. The resulting integers expressed as prime factorisations are as follows:

n	matchings
1	2
2	$2 \cdot 3^3$
3	$2^3 \cdot 3^3 \cdot 5$
4	$2^5 \cdot 5^7$
5	$2^2 \cdot 5^7 \cdot 7^5$
6	$2^8 \cdot 3^3 \cdot 5 \cdot 7^{11}$
7	$2^{13} \cdot 3^9 \cdot 7^{11} \cdot 11$
8	$2^{13} \cdot 3^{18} \cdot 7^5 \cdot 11^7$
9	$2^8 \cdot 3^{18} \cdot 11^{13} \cdot 13^5$
10	$2^2 \cdot 3^9 \cdot 11^{19} \cdot 13^{11}$
11	$2^{10} \cdot 3^3 \cdot 11^{19} \cdot 13^{17} \cdot 17$
12	$2^{16} \cdot 11^{13} \cdot 13^{23} \cdot 17^7$

For $n = 12$, the number of coverings is approximately 3.8×10^{52}. It is therefore highly suspicious that its highest prime factor is only 17. Numbers with this property of being large by comparison with their largest prime factor are often described as *smooth* or *round*:

The definition of roundness is not precise, since it is not intended for use as a technical term. Its vagueness is intended to capture the uncertainties and the suspense of formula-hunting, and the debatable issue of whether the occurrence of a single larger-than-expected prime factor rules out the existence of a product formula. (Propp 1999: 260)

You can also see that the exponents of the prime factors appear to occur in patterns (e.g. exponents of 3 are 0, 3, 3, 0, 0, 3, 9, 18, 18, 9, 3, 0, . . .). The appearance of round numbers right the way down this list and the exponent patterns will set the mathematician thinking that some neat formula is lurking about the data. As Propp remarks:

Most of the formulas that have been discovered express the number of matchings of a graph as a product of many comparatively small factors. Even before one has conjectured (let alone proved) such a formula, one can frequently infer its existence from the fact that the number of matchings has only small prime factors. (Propp 1999: 260)

Of course it is no easy matter to pass from this data to a formula. So to help in the search for such a formula, programs have been written. One such program is RATE [from the German verb 'to guess'] written by Krattenthaler.[2] With this resource and others, including Superseeker, the electronic Encyclopaedia of Integer Sequences, Krattenthaler reckons that when searching for a general formula for members of a sequence of integers 'guessing can be largely automated' (1999: 51).

RATE's deepest search involves looking for an analytic expression for the ratio of ratios of successive terms $(a_{n+2} \cdot a_n / (a_{n+1})^2)$. For example, if RATE is given the eight integers (1, 2, 7, 42, 429, 7436, 218348, 10850216) it will output

$$\prod_{r=1}^{n-1} 2 \left(\prod_{s=1}^{r-1} \frac{3(2+3s)(4+3s)}{4(1+2s)(3+2s)} \right).$$

With a little work this expression may be rewritten

$$\prod_{r=0}^{n-1} \frac{(1+3r)!}{(n+r)!},$$

the formula for the number of alternating sign matrices, i.e., matrices of 0s, 1s and −1s in which the entries in each row or column sum to 1 and the non-zero entries on each row or column alternate in sign, a result proved as recently as 1995.

---

[2] See http://radon.mat.univie.ac.at/People/kratt/rate/rate.html.

But even when you have a formula, your work is not finished. There is always the thought that a deep reason lies behind it. For example, as Necker cube-type optical illusions suggest, hexagons tiled with lozenges may be viewed as a two-dimensional projection of stacks of cubes. For the working mathematician, this may well evoke the notion of Young diagrams, a way of classifying representations of permutation groups. We are also making contact here with the study of integrable lattice models initiated by Onsager and Ising, ideas of which were used in a proof of the alternating sign matrix conjecture. We know what extraordinarily rich connections this has to mathematics. As Propp puts it:

> The deeper significance of these formulas is not clear. Some of them are related to results in representation theory or the theory of symmetric functions, but others seem to be self-contained combinatorial puzzles. Much of the motivation for this branch of research lies in the fact that we are unable to predict ahead of time which enumerative problems lead to beautiful formulas and which do not: each new positive result seems like an undeserved windfall. (Propp 1999: 256)

But even those formulas which 'seem to be self-contained combinatorial puzzles' usually possess what Propp (1999: 286) calls some 'gratuitous symmetry'. By this he means the property of a formula being invariant under the replacement of n by some linear expression in n, say, $-n$, which interpreted in terms of the tiled grid makes no sense.

The computer assisted part of this 'fishing' for combinatorial formulas is simple enough that undergraduates can be involved in the discovery of novel results, something rather rare in mathematics. There is a significantly inductive flavour to this work and yet the background theoretical framework is considerable, as evidenced by the facts that the procedure is highly specific to graph matching problems and that the formula guessing algorithm works within a tight range. Real numbers appear more frequently as mathematical data. Let us turn now to see how to automate the discovery of conjectures about relationships between sets of real numbers.

3.3 NUMERICAL IDENTITIES AND ENUMERATION

A new identity for π has recently been discovered:

$$\pi = \sum_{k=0}^{\infty} \frac{1}{16^k} \left[\frac{4}{8k+1} - \frac{2}{8k+4} - \frac{1}{8k+5} - \frac{1}{8k+6} \right].$$

This remarkable formula offers us a way to calculate a digit of the hexadecimal (base 16) expansion of π without calculating the previous digits.

You can now calculate on your own personal computer in around 30 seconds that the millionth hexadecimal place of π is 2. The identity may be proved by elementary calculus, but it was found by 'a combination of inspired guessing and extensive searching using the PSLQ integer relation algorithm.' (Bailey *et al.* 1997: 905).

This PSLQ algorithm is the result of a long-standing quest to find a way of determining whether for a set of n real numbers, x_i, there are integers a_i not all zero and not exceeding a specified size such that $\Sigma a_i x_i = 0$. One use of such an algorithm is to discover whether a given real number is algebraic, that is, satisfies a polynomial with integer coefficients. To see whether there is such a polynomial of degree less than n with coefficients not too large for some real α, PSLQ is set to search for an integer dependence between powers of α up to n ($1, \alpha, \alpha^2, \ldots, \alpha^n$).

During the running of the algorithm, y, a vector of real numbers associated to $\underline{x}$, is defined. While the search is proceeding the ratio between the largest and smallest entries of y is usually at most two or three orders of magnitude. The program continues either until the point is reached when one knows that if there is some integer relation then some of the a_i must exceed a large pre-set bound, or else the smallest entry of y suddenly becomes extremely small, providing very strong evidence that an integer relation has been found. An example of the former case was a search conducted to find a similar formula for π but with '10' instead of '16', so as to allow the decimal places of π to be calculated from a given place without knowledge of previous digits. Using PSLQ, we now know that no formula of similar simplicity can exist. In the latter case, when success occurs, Borwein and Broadhurst describe the final ratio between the smallest and largest entry of y as a 'confidence level', in the sense that, bearing in mind rounding errors, it would have been extremely unlikely for the algorithm to have behaved like this merely due to chance.

We have here then an example of a highly specific algorithm aiming to extract as much information as possible from a real number or set of real numbers, by working to many thousands of decimal (or some other base) places. High-precision arithmetic makes this possible, but even so, one clearly cannot throw too large a set of real numbers at the device, for then the discovery of apparent integer relations will be inevitable. The reader may consult Websites where vast numbers of mathematical constants are listed.[3] Aimless searching for relations between these cannot work. The

---

[3] See, for instance, Plouffe's inverter at http://www.lacim.uqam.ca/pi/ where 200 million constants may be consulted.

'inspired guessing' mentioned above must act to restrict severely the range of candidate numbers appearing in the formulas in something more closely resembling the mechanical falsification of chemical compound screening mentioned in section 3.1 above.

Moreover, as in the case in section 3.2, we should observe that the discovery of the formula for π is not in itself a fundamental piece of mathematics unless one finds richer connections to the identity. An indication that this may happen is a suggestion by researchers in the field that it may be possible to relate the production of digits of π by the discovered formula to the output of a dynamical system, with the further prospect of demonstrating the normality of π, i.e., the property that any given sequence of n digits occurs with frequency 10^{-n}. Let us turn now to the least specific algorithm to be considered in this chapter.

3.4 INDUCTIVE LOGIC PROGRAMMING

ILP has been defined by Stephen Muggleton, one of its leading exponents, as the intersection of machine learning and logic programming. Readers will perhaps be familiar with the first of these terms. While much activity in machine learning is aimed merely at constructing classification devices, ILP aspires to do more. When an ILP device is given background knowledge B, expressed as a set of predicate definitions, positive examples E^+ and negative examples E^-, it should construct a logic formula H such that:
1. all the examples in E^+ can be logically derived from $B \land H$, and
2. no negative example in E^- can be logically derived from $B \land H$.
Background knowledge, examples and hypotheses are represented as *logic programs*. This latter concept may be new to the reader, so let us see a simple example of what we may expect from PROGOL, one of the most advanced ILP algorithms.

The name PROGOL is derived from Prolog, a computer language based on the predicate calculus, or rather that portion of it expressible in Horn clauses.[4] Given data about the result of appending two lists of constants, we would like PROGOL to be able to generate a definition of the append operation for us. First, we need to provide background information about the types involved: lists and constants. We input

```
list ([]).
list ([H|T]) :- const(H), list(T).
```

---

[4] The 'GOL' part is the reverse of the 'log' of Prolog, an allusion to the thought that induction is the reverse of deduction.

This recursive definition tells PROGOL that a list is either empty or has a constant at its head followed by another list (read ': -' as a backwards implication '←', and ',' as conjunction). Next we need to let PROGOL know some candidates for list membership,

const(a). const(b). const(c). ...

Finally, we need a sample of examples of the form

append([a, b, c], [d, e], [a, b, c, d, e]).

We may also give some negative examples, such as

:- append ([a, b], [d, e], [a, b, e, d]).

or perhaps include a clause to state that append is a function.

What we hope to acquire as output is a recursive definition of append such as:

append([], A, A).
append([A|B], C, [A|D]):- append(B, C, D).

To do so we must provide templates for the clauses appearing in the output. These 'mode declarations' tell PROGOL to search for generalisations of the data of a certain form, e.g., one might have

:- modeb(1, append(+list,+list,-list))?

if one thinks append(A, B, C), with A and B as input lists and C as output list, might well appear in the body (right-hand side) of a suitable clause. From the language of the examples and background knowledge, the algorithm produces hypotheses which are constrained by the mode declarations and which are entailed by the background knowledge and the negations of the positive examples. Promising hypotheses are those that combine brevity with the ability to account for as many as possible of the positive examples, but no negative examples, in the context of the background knowledge. If PROGOL is carrying out positive-only learning, it will randomly generate potentially negative clauses.

We can easily express PROGOL's results above in ordinary language. Appending the empty list to any list leaves it unchanged. If D is the list formed by appending B to C, then appending B with an additional constant at its head to C will result in D with that additional constant at its head. The capacity of an ILP program to produce hypotheses permitting this kind of translation is frequently cited as a key advantage over a statistical classification device, which in this case would merely be able to give you a numerical value expressing the extent to which it was certain that one list was the result of two others being appended. Indeed, anything PROGOL learns may be translated into natural language, with the possibility that an expert in the domain can make sense of it. In the case referred to in the introduction to this chapter, the law found to classify the mutagenicity of compounds was as follows:

A compound is highly mutagenic if it has (1) a LUMO [the energy of the Lowest Unoccupied Molecular Orbit] value ≤ -1.937; or (2) a LUMO value ≤ -1.570 and a carbon atom merging six-membered aromatic rings; or (3) a LUMO value ≤ -1.176 and an aryl-aryl bond between benzene rings; or (4) an aliphatic carbon with partial charge ≤ -0.022. (Muggleton 1999: 7)

This uniformity of the treatment of background knowledge, examples, and output is cited as one of ILP's strengths.

So where now to turn for suitable mathematical data to let our ILP program loose on? We have seen two examples of PROGOL at work, a textbook example appearing as a manual illustration, and a more serious piece of work in biochemistry. In the biochemistry case, one has individual compounds each with a collection of relevant data. PROGOL is trying to generate a generalisation from a single case, according to advice as to what form it should be looking for. Each time it generates a candidate hypothesis PROGOL then see how many examples it covers. We appear to be in a similar situation with the integer relation search as an expression is associated with the many places of its decimal expansion. However, since PROGOL does not contain any particularly sophisticated ways of treating numerical data, we cannot hope to compete with the extremely specialised PSLQ algorithm. Similarly, nor should we expect PROGOL to be able to compete with RATE on integer sequences. One might suppose the logic programming approach to work better on mathematical data where there are expected connections between different pieces of data to allow the kind of recursive output of the 'append' example. With all these considerations in mind I opted for data concerning the homotopy groups of different dimensional spheres.

3.5 THE PROBLEM OF THE HOMOTOPY GROUPS OF THE SPHERES

Homotopy theory is a branch of algebraic topology. The algebraic topologist's principal aim is to find refined, yet tractable, ways to associate algebraic invariants and mappings to topological spaces and continuous mappings.

Homotopy theory considers some of the most subtle invariants of this kind. One of its central constructions is Map(X, Y), the set of ways a space X may be mapped continuously into another Y,[5] where one chooses not to distinguish between mappings whose images are merely continuous deformations of each other. In many situations it may be possible to equip Map(X, Y) with some extra structure. For instance, in the cases we shall consider here it is possible to compose any two mappings to result in a third, and indeed so as to form a commutative group.

One approach to the analysis of topological spaces views the spheres as the simplest building blocks of spaces.[6] It seems reasonable then to use these spheres as probes to detect the topological structure of a target space. Like a fisherman we cast circles, balls, and their higher-dimensional equivalents, hoping to catch something within the target. So, define $\pi_n(Y) \equiv$ Map(S^n, Y) to be the nth homotopy group of Y. An obvious choice for Y then is another sphere, an m-sphere, say.

For $n = 1$ and $m > 1$, $\pi_1(S^m) = 0$, the trivial group. This is simple to see when $m = 2$. Then any loop on the surface of a ball may be contracted to a point. But we do pick up some homotopy when $m = 1$. $\pi_1(S^1)$ measures the ways of mapping a circle into another and is equivalent to the integers under addition. This you may be able to imagine by wrapping string around your finger a number of times. Going around three times and then twice more corresponds to adding 3 and 2. Reversing the direction of the two winds corresponds to the subtraction, 3 minus 2.

This ability to wrap an n-sphere around another n-sphere continues in higher dimensions. With the ordinary sphere, for instance, for a natural number d, one can imagine cutting a latex globe down the Greenwich Meridian and pulling on one side of the cut to wrap it round the globe d times, before stitching it back up along the Meridian. This is all the homotopy to be had using these dimensions: for all $n > 0$, $\pi_n(S^n) = \mathbf{Z}$, where Z stands for the group formed by the integers under addition.

Now, for n less than m, there is enough room in the target m-sphere to contract any image of the n-sphere to a point. In other words, for all $n < m$, $\pi_n(S^m) = 0$. You might think conversely that if $n > m$, there would not be enough room to map an n-sphere to hook onto an m-sphere, but you would be very wrong. It is true that $\pi_2(S^1) = 0$, balls cannot

---

[5] One also requires that a base point be chosen for each of X and Y, and that this be preserved by the mappings.

[6] n-spheres are defined for all values of n as homeomorphic to (that is, in continuous bijection with) the subspace of a $(n + 1)$-dimensional Euclidean space of points at unit distance from the origin.

be mapped usefully onto circles, but in the 1930s Hopf discovered that $\pi_3(S^2) = \mathbf{Z}$.[7] On a good day you might be able picture this so-called *Hopf fibration* in your mind. Things start to get much harder as m and n increase, however. Even the fact that $\pi_4(S^2)$ is the two-element group seems to be beyond most people's imagination.

Whereas for some Y we know $\pi_n(Y)$ for all n, surprisingly perhaps, not all $\pi_n(S^m)$ are known. Indeed, these groups are notoriously difficult to calculate. Even for $m = 2$, the ordinary sphere, not all homotopy groups are known. But still, for a large number of pairs (n, m), these groups have been discovered. For instance, strange though it may seem to think about this, it is known that the ways of mapping a 33-sphere into a 14-sphere form a group of order 2112. Some readers may at this point be wondering why on earth we should want to know such things. Well, those who prefer their mathematics to have some relevance to the world may be comforted to learn that faults in nematic crystals may be classified by knowledge of $\pi_n(S^m)$ for low values of m and n. As for higher values of m and n, there is a thought around that $\pi_{11}(S^8) = C_{24}$ has relevance to M-theory, the next big thing for string theorists. More generally, the techniques devised in homotopy find use in algebraic topology as a whole and from there in other branches of mathematics and mathematical physics.

To sum up what we have learned so far, for any two given positive integers, n and m, we have a finitely generated Abelian group, $\pi_n(S^m)$. If these groups were entered on a grid of n against m, we would have $\mathbf{Z}$ all the way along the main diagonal, 0 in all positions to one side of it, and a complicated array of groups on the other. It turns out that these groups are by no means unrelated. Indeed, there are many subtle interrelations between the groups. One of the most important of these is known as stabilisation. It concerns homotopy groups for a fixed difference k between m and n, i.e., entries along a diagonal. Then, for n less than $k + 2$ the groups will generally vary, but thereafter all groups are identical. So, for example, for $k = 3$, $\pi_8(S^5) = \pi_9(S^6) = \pi_{10}(S^7) = \cdots = C_{24}$, the cyclic group of order 24. Part of the work of calculating these groups involves understanding how elements are preserved as one passes along the diagonal. These so-called stable homotopy groups are easier to calculate, but still there is no general formula. It is possible to split these groups, rather like we do with numbers, into their prime components. Some dedicated souls have managed to calculate

---

[7] For those who care to know such things, a map from the 3-sphere to the 2-sphere which generates this group may be represented algebraically as a map from the unit quaternions (isomorphic to the 3-sphere) to the Riemann sphere (the complex plane compactified by a point at infinity).

the 5-component of the stable homotopy groups up to $k = 1000$, giving us information about the ways the 2002-sphere maps into the 1002-sphere!

In a perfect world we would find an algorithm which when two integers and a prime (n, m and p) are entered as input, sends back the p-component of the nth homotopy group of the m-sphere. Mathematicians are far from that point, however. Instead, there is an extraordinary range of techniques used in the calculation of the groups. For example, from the existence of a fibration, resembling the Hopf fibration, except this time the total space is the 7-sphere with fibres homeomorphic to the 3-sphere and base space the 4-sphere, we may derive the relation $\pi_q(S^4) \equiv \pi_q(S^7) \oplus \pi_{q-1}(S^3)$. This is a simple general relation, but as n and m increase more intricate tricks must be used. Out of this motley of ways of calculating the groups, might one hope perhaps to find some unnoticed regularities?

We need to present the data to PROGOL. Now, finitely generated Abelian groups are sums of a number of copies of the infinite cyclic group and a finite Abelian group. A complete set of invariants for a finite Abelian group is given by a list of integers greater than 1, each of which is a multiple of its successor. For example, $\pi_{33}(S^{14}) \equiv C_2 \oplus C_4 \oplus C_{264}$. However, this representation is not well suited for my purposes, since it does not fit well with the operation of addition on Abelian groups. Better then to break these groups into their p-components for prime p. Very much like the factorisation of natural numbers, this decomposition is unique, however to a given prime power there correspond different Abelian groups. However, an Abelian group whose order is the power of a prime, p, is uniquely expressible as a sum of cyclic groups each of order some power of p. Thus we may express the Abelian group $C_2 \oplus C_4 \oplus C_8$ as [1, 1, 1], while, say, $C_2 \oplus C_2 \oplus C_8 \oplus C_{16}$ may be expressed as [2, 0, 1, 1].

We enter positive examples of the form h(2, 33, 14, [1, 1, 1]). This says that the 2-component of $\pi_{33}(S^{14})$ is the 64-element group $C_2 \oplus C_4 \oplus C_8$. We do not need to enter negative examples as we can also specify in the background knowledge that

```
:- h(A, B, C, D), h(A, B, C, E), D ≠ E.
```

This tells PROGOL that the A-component of the Bth homotopy group of the C-sphere is uniquely specified. Other background knowledge will include the fact that the homotopy for B greater than A is trivial:

```
h(2, A, B, []) :- A ≤ B.
```

As with all machine learning algorithms, combinatorial explosion is always a threat. As I mentioned above, PROGOL 4.4 escapes from this threat by the use of mode declarations and background knowledge. These delimit the range of hypotheses the computer can look for and so reduces enormously the lattice in which it searches. In this case we know that there are relations expressing one group as the sum of two other groups. As we have represented the p-component of an Abelian group for prime p as a list, we need to define the addition of two lists to correspond to the addition of the groups. This is easily done as a recursive definition in PROLOG:

```
list_add([H1|T1],[H2|T2],[H3|T3]):- H3 is H1 + H2,
list_add(T1, T2, T3).
list_add([], L, L).
list_add([H|T],[],[H|T]).
```

As for mode declarations, we know that a target law is $\pi_q(S^4) \equiv \pi_q(S^7) \oplus \pi_{q-1}(S^3)$. This holds, in particular for the 2-component, which PROGOL would represent as:

```
h(2, A, 4, B):- h(2, A, 7, C), h(2, D, 3, E),
add_list(C, E, B), succ(A, D).
```

This suggests we guide PROGOL by making mode declarations such as:

```
:- modeh (1, h(#nat,+nat,#nat,-list))?
```

and

```
:- modeb (1, add_list(+list,+list,-list))?
```

3.6 RESULTS AND DIAGNOSIS

So, how did I fare with PROGOL on this homotopy data? Well, I have to confess to achieving little success. With some considerable prompting it did manage to find

```
h(2, A, 4, B) :- h(2, A, 7, C), h(2, D, 3, E),
add_list(C, E, B), succ(A, D).
```

but this did require some generous hints in the form of mode declarations, and I was unable to come up with anything else. Of course, one philosopher's failure to make a machine learning system generate any interesting mathematical conjectures in one particular problem situation provides very little evidence that that machine learning systems are poorly adapted for mathematical reasoning at the inductive end of the scale. It may well be that even the self-same system I used could be made to perform far more effectively in the hands of an expert working on this very problem. However, as further evidence that one should not expect anything significant from such a computer assisted approach I can report to you the outright incredulity on the part of the professional topologists I consulted as to the prospects for PROGOL's success. As John Earman comments in his (1992), extreme implausibility of this kind is not to be ignored by the philosopher. He illustrates the thought that there is 'sharp consensus' about 'unnecessary experiments' with a cartoon showing, alongside other frames, one in which a scientist stands on the rim of a volcano testing whether a paper towel can stop it erupting (Earman 1992: 159). What lies behind the algebraic topologists' incredulity here? Why is PROGOL as likely to extract important unknown relations from homotopy data as a paper towel is to protect the residents of Ercolano from Vesuvius?

One way of seeing what is going on here is to think through an idea expressed by Ronald Brown, an algebraic topologist whose interests include the computerisation of algebraic calculations. Brown claims that:

Mathematics often proceeds *Geometry* → *Algebra* → *Computation*. (Brown 1992: 243)

and in a recent unpublished paper he has elaborated this schema to the following:

geometry → underlying problems → algebra → algorithms → computer implementation.

There is an important idea here. Very many key constructions in algebra have arisen out of a desire to penetrate to the heart of geometric or topological problems. For example, we can certainly think of Descartes's analytic geometry in this way. But, in such a process it is not merely a question of devising as powerful a form of algebraic representation as possible. There is a balance to be struck between there being sufficient subtlety in the algebra to capture salient features of the geometry, without sacrificing tractability. The C^*-algebra of continuous complex functions on a topological space completely specifies that space, in that the latter is recoverable up to homeomorphism from the algebra. However, nothing is gained for topology since the algebra is no more tractable than the topology.

It should be noted that mathematicians also work in the opposite direction, from algebra to geometry. For instance, having observed the aforementioned equivalence of commutative C^*-algebras and topological spaces, mathematicians set themselves the task of finding the 'spaces' giving rise to non-commutative C*-algebras. This search for the leftmost portion of an instance of Brown's schema was one of the founding steps of noncommutative geometry. In general terms, geometry and algebra are more intricately involved with each other than the schema suggests, and we may complicate it further by observing, as we shall discuss in chapter 10, that it is possible to use topological diagrams to perform algebraic calculations.

What we are trying to do in the cases considered in this chapter is to pass back along the chain of the schema from the results of computations to gain new insight into the algebra, as well, possibly, as its algorithmic implementation, and from there potentially to the geometry and its underlying problems. In the case of graph matchings, we possess an efficient algorithm to pass from large matrices to their determinants, allowing us to enumerate the large number of distinct matchings. In these calculations, we have not had to rely on deeper knowledge of the geometry–algebra relation, namely, the graph matching–representation theory relation, but we do know enough about the algebraic situation to expect certain sorts of numerical relation to hold among the data, which we can then search for algorithmically. We would hope then that the discovered formula would provide some new insight into representation theory.

The discovered formula for π is as yet a rather peripheral result. It can be proved easily enough by integration of simple functions, but it is unlikely one would have thought to do so before the formula was known. High-precision arithmetic is used to evaluate certain sums, producing a surfeit of data. Then a knowledge of the kinds of relation in which to expect π to occur greatly constrains the search. The discovered algebraic formula may then be linked to the geometry of a certain dynamical system.

In both of these cases, computers programmed with general purpose algorithms are being used to generate novel information which is somewhat independent of what we know of the underlying algebraic situation, which in turn provides a way of sharply constraining the search for a conjecture. In both cases, I would incline to view them on the side of falsification rather than induction.

As for our final example, homotopy theory is one of the subtlest forms of algebraic topology, with the unfortunate consequence that it is very hard to perform calculations there. So, other more tractable theories (ordinary cohomology, K-theory, complex cobordism theory, BP-theory, etc.) are used which can capture some of the homotopy. As we shall see in chapters 9 and 10, there are those, Brown among them, who believe it is possible to develop the 'right' algebra for homotopy theory using higher-order groupoids, a radical departure from the algebra we have grown to love, while retaining some computational tractability.

These considerations point us to two sources for the problems PROGOL faced: (i) the large amount of extremely intricate algebraic and geometric theoretisation which allowed the groups to be calculated in the first place means that data is not acquired by independent sources; (ii) the predicate calculus is not well suited to encoding this algebraic and geometric background knowledge, nor to providing a framework in which an algorithm linking algebra to computation may be approached. Let us take these in turn.

Independent sources

If you are interested in the drug toxicity of a group of compounds, you put two teams onto the task of collecting data. They will typically have very different skills. While the first carries out chemical analysis on each compound to discover its molecular structure and properties, the second injects each compound into samples of unfortunate animals to see how long they survive, or how serious are the compound's effects. In this way PROGOL is being given data from two independent sources. Any relations discovered between these two types of data constitute new information. With the homotopy data, on the other hand, a lot of high-level theory, of precisely the type sought, has already been used to generate the data in the first place. To get a flavour of the overwhelming amount of theoretical machinery devised for homotopy theory, listen to an appreciative reviewer's summary of the first chapter of Ravenel (1986), a book in which it is explained how to go about calculating homotopy groups:

One of the nice features of this book is Chapter 1, 'An introduction to the homotopy groups of spheres'. It begins with a quick historical survey, starting with the Hurewicz and Freudenthal theorems and leading, via the Hopf map, to the Serre finiteness theorem, the Nishida nilpotence theorem, and the exponent theorem of Cohen, Moore, and the reviewer. Then results relating to the special orthogonal group are described, for example, Bott periodicity and the image of J. The history of computing homotopy groups is illustrated by a brief discussion of the Cartan–Serre method of killing homotopy groups and of its descendent, the classical Adams spectral sequence. Some of the triumphs of this spectral sequence, or, more precisely, of the secondary cohomology operations related to it, are indicated; for example, the solutions to the classical and mod p Hopf invariant one problems. At this point, the author makes the transition to the main subject matter of this book by describing the complex cobordism ring, formal group laws, and the Adams–Novikov spectral sequence. The applications of this and related techniques to the existence of infinite families of elements in the stable homotopy groups of spheres are then indicated. Next, the author replaces cobordism by the more tractable BP-theory and introduces the chromatic spectral sequence. Chapter 1 closes with a discussion of the way in which the unstable homotopy groups of spheres relate to the vector field theorem, the Kervaire invariant, and the Segal conjecture. Present in this discussion are James periodicity, the *EHP* sequences of James and Toda, and the Kahn–Priddy theorem. The description of Mahowald's work on the stable *EHP* spectral sequence is likely to be of special value to the experts. It should be clear that a reader of Chapter 1 can come away with some understanding of a substantial portion of current homotopy theory. (Neisendorfer, review of Ravenel 1986 on MathSciNet)

What we do not have here, at present at least, is the prospect of performing the homotopy calculations in terms of a more general purpose algorithm.

Aside from the general purpose algorithms, one might imagine cases in some situations where diagrams could provide an alternative source of knowledge, as James Brown (1999) sees their role. Unfortunately, direct pictorial data is extremely rare. Our visualising capacities are adequate to provide sufficient data to convince us of the truth of 'Vertices – Edges + Faces = 2' for sphere-like polyhedra, but cases like this are far from typical. In our case, the fact that the third and higher homotopy groups of the 2-sphere are the same as those of the 3-sphere can be 'seen' from the existence of the Hopf fibration, where the 3-sphere, which although three-dimensional cannot be embedded in three-dimensional space, is fibred over the 2-sphere. On a good day I can glimpse all this, but it cannot take us very far.

Two extreme strategies, then, are either to pack as much as possible of the data-generating theory into the background knowledge, but then the positive examples, E^+, are already consequences, or else to feed in very little by way of background knowledge, hoping to pick up regularities in the data independent, as it were, of the motley of high-level theory. This latter strategy strikes the experts as unworkable – surface regularities are rare and in any case well understood. This leaves the possibility of an intermediate strategy, but one would imagine that it would suffer from the difficulties of the extremes to the extent that it resembles them.

Language

Homotopy theory is an unfinished story. The possibility for higher homotopy groups was recognised in the 1930s, yet when it was shown that they are all Abelian, it was enough to persuade leading topologists that this was not the way forward. In hindsight this seems a drastic reaction. After all, Abelian groups give you some information. But the reaction was based on some correct intuition: that there ought to be a more comprehensive way of capturing higher-dimensional information. As we shall see in chapter 9, there are indications that different types of algebraic object, known as higher-order groupoids, can achieve this. Even at the simplest stage, ways of calculating the fundamental group of the circle, $\pi_1(S^1)$, generally wander away from the simple attribution of algebraic object to topological object, while using groupoids instead of groups allows you to remain within this attribution. In higher-dimensional situations, multiple groupoids have been found to capture the topology algebraically. Furthermore, groupoid theory is now being welded onto the GAP computer algebra software package to allow for automated computations.

If all this were to go smoothly, then Brown's schema would present a very clear picture of the situation, but it is hard to see an ILP device contributing to this process. To fill the gap between algebra and computation requires efficient algorithms. But the logical representation of even the simplest algebraic calculation is extremely long-winded. For example, it has been estimated that to show that $(x + y)^2 = x^2 + 2xy + y^2$ it takes apparently over 70 steps in the natural deduction calculus (Kerber *et al.* 1998: 337). First-order logic is an 'algebra' primarily suited to finite sets, with properties and relations. Where it is happiest is dealing with a database of entities and their attributes. Nor would the availability of set theoretic language help the situation – it is for good reason that computer algebra systems do not rely on set theoretic reductions of their subject matter.

Once we have the right algebra and have encoded a suitable algorithmic implementation, the effects can be dramatic:

Algebraic geometers already got used to the possibility of computing the cohomology of coherent sheaves by hitting some keys on a keyboard, but questions pertaining to the topological cohomology of algebraic varieties (especially open or singular) are still regarded as belonging to some kind of 'Alaskan refuge' of contemplative thinking. The possibility of an 'industrial approach' to this area has a sobering effect, forcing one to ponder wider issues raised by the use of computers. (Kapranov 2001: 487)

Going backwards to discover the implementation from the data is impractical.

On the other hand, moving to the earlier links of Brown's schema, it is hard to think that PROGOL can help find the 'right' algebraic language since this presupposes considerable theoretical understanding of that domain. For one thing, PROGOL is treating the homotopy data as a collection of functions taking spheres as input and outputting groups. Mathematicians see the homotopy groups as functors (see appendix), where the mappings between the spheres, with their geometric interpretations, get represented by mappings between groups. It is hard to see how to encode effectively the existence of such relations between spheres. But this is only the beginning of mathematicians' understanding of the geometric situation. Homotopy groups admit a multiplication, something that can be seen from thinking in terms of spheres, and so form what is called a graded algebra. Encoding substantial background knowledge about spheres seems implausible.

3.7 CONCLUSION

In treating at length one attempt to use one system on one problem, it might be thought that any claim to generality had been lost, but I believe that the vividness of an allusive illustration of this kind is adequate recompense. A view of mathematics as tightly bound up with logic would tend to suggest that machine learning systems, especially those using the resources of the predicate calculus as mine did, should be very well suited to problem tasks in mathematics, or at least as well suited there as in the natural sciences. If not producing mathematics on their own, they should at least be of some considerable assistance to the research mathematician. This, however, is not the case. In fact, it is the data-rich fields such as biochemistry where there is the best prospect of success. This picture is repeated throughout automated reasoning in the sciences.

It could hardly be expected otherwise than that, as we proceed along a scale leading from a highly tailored algorithm plus considerable human intervention to a general purpose algorithm with little intervention, we achieve less. In the first of our examples, because of the data generation process, we saw significant results about graph matchings discovered without passing through deeper explanations from the theory of group representations, although their existence is to be expected. The algorithm used in the second case was less constrained, it can be used on any moderately small finite set of real numbers. But here there must be reasons to believe that integer relations might exist. Although the discovered formula suggests that the digits of π could be seen as generated by some pseudo-random process associated with a dynamical system, this played no part in the discovery. Finally, where a general algorithm with no specific ability to represent a particular kind of mathematical entity operates on data concerning that kind of entity, little is to be expected.

In some sense, this was an unfair task. There are, however, many reasons to attempt the impossible or near impossible deliberately. Whereas King Canute ordered the sea to retreat to reveal the ineffectual sycophancy of his advisors, we are told, here I wanted to show how algorithmic inductive activity should be seen as much less likely to be effective in concept-rich disciplines such as mathematics than in the data-rich disciplines of some of the natural sciences. A second goal was to establish further the inadequacy of the predicate calculus as a means of representing mathematics. First-order logic does not often bring out the grain of mathematics despite the hopes of Frege. As Tappenden (1995) shows, Frege had imagined that his new proto-predicate calculus, the *Begriffsschrift*, would allow mathematicians to carve out concepts fruitfully. That hope has only very partially been realised.

To the extent that ILP has achieved some phenomenological success in areas of natural science, picking up surface patterns in bodies of data, we can say that the predicate calculus is a suitable language in which to express generalities there. My sense is that the ILP paradigm will always find life harder in mathematics. Plenty of results in mathematics appear to be of the simple logical form, all S are M – for instance, all elliptic curves are modular. Lakatos (1976) showed us, however, that the difficult part is the construction of the definitions involved in S and M. If you know the composition of the conjecture to that extent, you have done much of the hard work and you will have considerable insight as to how to approach a proof. In the drug-toxicity case, on the other hand, this need not be the case.

This brings us to the end of our foray into the borderland between articial intelligence and mathematics. I hope that the reader is now inclined to think that philosophy can gain impetus by juxtaposing itself to novel disciplines. There is certainly much for practice-oriented philosophers of mathematics and computer scientists working on automated mathematical reasoning to talk about. Several of the thirty-three questions of Larry Wos are ones pondered by philosophers. Even if philosophers are sceptical of the present capabilities of computers, detailed arguments as to the reasons for their doubt should clarify important issues. Now, after two chapters of machine-based research at the periphery of mathematics, let us change gears and look at the way humans do mathematics, the kinds of human Hardy would have counted as 'real' mathematicians.

CHAPTER 4

The role of analogy in mathematics

> The enrapturing discoveries of our field systematically conceal, like footprints erased in the sand, the analogical train of thought that is the authentic life of mathematics.   (Rota in Kac *et al.* 1986: ix)

4.1 INTRODUCTION

When mathematicians talk informally about a theory, their metaphors and similes often reveal something about their conceptualisation which too often goes missing from the papers and monographs they write. The topologist Solomon Lefschetz describes, for instance, how back in the 1920s he 'planted the harpoon of algebraic topology into the body of the whale of algebraic geometry' (1971: 13). By this he means that at that time algebraic geometry encompassed a large amount of rather poorly structured theory. Algebraic topology was a newly created tool which Lefschetz could employ to capture and systematise the older theory. There is also an allusion to death in his metaphor which agrees well with mathematicians' use of this word to describe a theory which has been 'killed off' by being worked out.

   A more explicit similarity lies behind the following statement: 'The zeta function of a field is like the atom of physics . . . we will show how to split it via group theory' (Stark 1992: 366). While we have two objects which at some point in their careers were profitably split, revealing in each case important information about the relation of the whole to its parts, I imagine that when Stark wrote this he did not believe that his comparison could be pushed too far. Of course, there was a similarity in the fact that representation theory is used to factorise the general zeta function and is used in the construction of the quantum theory of the atom, yet it is unlikely that he believed there would be any meaningful equivalent of, say, the nucleus in the number theoretic case. Curiously, though, a powerful connection between these domains has since been found. The (scaled) spacings between the zeros of the Riemann zeta function and between the energy levels of a heavy nucleus turn out to share the same statistical distribution.

In this chapter I want to examine cases of quite precise structural similarity and so I shall talk of 'analogy' rather than 'metaphor'. To elaborate a vague idea relating two apparently distinct theories is often described in the most loving terms by mathematicians.[1] But this sometimes mysterious, yet common, use of concepts drawn from one area to illuminate another body of theory may be seen not only as a source of satisfaction for mathematicians, but also as a riddle begging philosophical treatment, what might be termed the 'unreasonable effectiveness of mathematics in mathematics':

> The glory of mathematics lies not so much in the fact that abstract theories do turn out to be useful in solving problems, but in that wonder of wonders, in the fact that a theory meant for one type of problem is often the only way of solving problems of entirely different kinds, problems for which the theory was not intended. No philosophy of mathematics can be excused from explaining such occurrences. (Rota 1991: 448)

Examples are never far from hand: Hermitian metrics from differential geometry are used in number theory, knots embedded in three-dimensional manifolds are usefully likened to prime ideals in the spectrum of a ring of algebraic integers in arithmetic topology, and so on.

In the days when discovery and justification were held to be quite separate processes it might have been argued that the use of analogy pertains merely to the *discovery* or *invention* of mathematical theories, but has little to do with the *justification* of mathematical results. Often the sense of mystery accompanying a perceived analogy is dispelled when the structural similarity is explained away in some larger theory, thereby revealing to us how it all works. But we can view matters differently. Analogies do not just act as cognitive aids to the discovery and learning of mathematics and so relate solely to the psychology of mathematical understanding. Important though this is in itself, they also play a central role in determining the direction mathematics has taken throughout its history by constituting what is deemed to be interesting theory. This role was described by Poincaré when he remarked that:

> the mathematical facts worthy of being studied are those which, by their analogy with other facts, are capable of leading us to the knowledge of a mathematical law, just as experimental facts lead us to the knowledge of a physical law. They are

---

[1] Rather incredibly, van Bendegem can claim 'both analogy and metaphor are . . . disliked by mathematicians' (van Bendegem 2000: 106).

those which reveal to us unsuspected kinship between other facts, long known, but wrongly believed to be strangers to one another. (Quoted in Atiyah 1978: 76)

Those wishing to deflate the inherent-structurist interpretation of this claim may choose to argue that the linking of apparent strangers is a relatively easy thing to achieve, so that constructions from more or less any domain could be used to develop any other if pushed hard enough. Indeed, so prevalent is the application of one theory to another that the suspicion must arise that this is so.[2] Then the analogies that are accepted have nothing more to recommend themselves than that they are accepted. We may believe, for example, that it is simply *natural* for functional analysts to take a function to be a 'point' belonging to a space, allowing geometric notions to provide them with the idea of nearness between functions. And we may well imagine that had Riemann, Ascoli, and Arzelà not started this way of thinking, then somebody else would have done so. But how much freedom was there to choose to make this analogy in the first place, were there plenty of other choices to be made, and once a function *was* seen as a point how much freedom was there to develop functional analysis? These are the kinds of question the philosophy of real mathematics needs to address.

Mathematicians tend to oppose 'nominalist' and 'contingentist' answers to these questions. Although through the twentieth century the technique of projecting parts of mathematics onto others has become totally unexceptional, leading Sir Michael Atiyah to claim that mathematics *is* the 'science of analogy' (Atiyah 1976: 220),[3] still powerful applications require the very great intellectual effort of uncovering profound analogies. The kind of intricate interaction that exists between the branches of number theory, algebra, geometry, topology and analysis, is really not so easy to achieve, but is widespread since it is precisely what mathematics is about:

[t]his interaction is, in my view, not simply an occasional interesting accident, but rather it is of the essence of mathematics. Finding analogies between different phenomena and developing techniques to exploit these analogies is the basic mathematical approach to the physical world. It is therefore hardly surprising that it should also figure prominently internally within mathematics itself. (Atiyah 1978: 75–6)

Contingentists might challenge this position by demonstrating how interaction of such a kind came to be seen as 'the essence of mathematics' and

---

[2] See Ruelle (1988) for thoughts along these lines.

[3] Atiyah goes on to suggest that 'the widespread applicability of mathematics in the natural sciences, which has intrigued all mathematicians of a philosophical bent, arises from the fundamental rôle which comparisons play in the mental process we refer to as "understanding"'.

how this image is maintained. While it is very clear that the explicit use of analogy has increased through time, this might be better thought of as an achievement, rather than the discovery of an essence.

In the space of a chapter I cannot do justice to these matters, but one of the central points I shall be making does bears on them. I claim that the development of the analogy between function and number initiated by Dedekind and Weber in the late nineteenth century should be seen as a watershed, marking a new level of sophistication in the use of analogy and the appearance of a new attitude to the possibilities of relationships holding between constructions from apparently different domains. The axiomatics of Hilbert, the algebra of Noether, and the category theory of Eilenberg and Mac Lane may be seen as increasingly sophisticated ways of capturing these relationships.

I shall treat the Dedekind–Weber case in sections 4.3 and 4.4, after reviewing other contributions to the use of analogy in mathematics. Since the idea of partner theories displaying structure similarity runs counter to the reductionist spirit of much of recent philosophy of mathematics, the literature on this subject is not extensive. I shall discuss in section 4.2 the ideas of the mathematicians Saunders Mac Lane and George Pólya and those of the philosopher Emily Grosholz and the sociologist of science Andrew Pickering.

4.2 THOUGHTS ON ANALOGY

For Mac Lane (1986: ch. X) analogy, along with the study of examples, the analysis of proofs, shifts of attention and the search for invariant formulation, are the most important methods we employ in our effort to understand mathematics. In what one might consider as the standard contemporary account of mathematical analogy, he tells us that an analogy perceived to exist between two or more theories suggests the possibility of generalising constructions in one of these theories by transferring them to the other and furthermore implies the existence of a common structure to be captured by an abstraction.

Analogy for Pólya forms one of the methods of plausible reasoning in mathematical discovery (Pólya 1954a: ch. II), and is based on the hope that a common ground exists between two domains. Pólya gives some examples of how we use analogy to pass from plane geometry to solid geometry. For instance, the tetrahedron is the analogue of the triangle because it is the solid figure with the least number of plane faces while the triangle is the plane figure with the least number of straight edges. Here, the source and target domains of the analogy resemble each other very closely. In his most detailed case study, on the other hand, there is a greater distance between the domains. This example concerns Euler's extension of the use of factorisation from polynomials to infinite series in response to the problem posed by Jacques Bernoulli of finding the sum of the reciprocal of the squares of the natural numbers. Euler's creative step was to treat $\sin x$ like a polynomial, reasoning that there was a chance that this would work in view of the shape of the graph of this function, the sine wave and its series expansion, which converges for all values of x. He had to hope that the fact that $\sin x$ has infinitely many roots would not matter. These roots are $x = 0, \pm\pi, \pm2\pi, \pm3\pi, \ldots$. Now, polynomials may be factorised as a constant multiplied by the product of factors of the form $(1 - x/\text{root})$. The analogy suggests then that, since $\sin x/x \to 1$ as $x \to 0$,

$$\frac{\sin x}{x} = \left(1 - \frac{x^2}{\pi^2}\right)\left(1 - \frac{x^2}{4\pi^2}\right)\left(1 - \frac{x^2}{9\pi^2}\right)\cdots$$

Using the series expansion of the left-hand side as $1 - x^2/3! + x^4/5! - x^6/7! + \cdots$, Euler compared coefficients of x^2 on each side to arrive at the result $\Sigma 1/r^2 = \pi^2/6$, which he then confirmed numerically to several decimal places. Here then the complex schema of 'factorisation to solve equation', or better its reverse 'construction of equation from solutions', has been transferred from the domain of polynomials to the domain of infinite series.

There is something almost Piagetian happening here. Piaget proposed the mechanisms of assimilation and accommodation as key to a child's intellectual development. Assimilation occurs when a schema is applied to new situations. This often leads to an imbalance, a lack of equilibrium, which is resolved by accommodation, the process of modifying the schema. Creativity for Piaget occurs through 'spontaneous assimilation'. In our example we find Euler creatively applying an existing schema to the domain of infinite series. But then what are the limits to the transferral? If the process works for $\sin x/x$, we might expect it to do so for similar functions. How similar? $\tan x/x$ has the same zeros as $\sin x/x$, and yet its series expansion differs in its second term, so this technique fails there. Might the failure be due to the function being undefined at infinitely many values of x? The function e^x, on the other hand, is defined everywhere, but it has no zeros. An interesting problem thus emerges of discovering to which series the technique can be applied. This process, which we may liken to accommodation, is frequently very fruitful, and in the case before us led to a 'common ground' in the form of a general theory of factorising entire functions, complex functions holomorphic throughout the complex plane.

Language similar to Piaget's is used by Pickering (1997) in his account of Hamilton's search for an analogue of the complex numbers in three dimensions, where he talks of a 'dialectic of resistance and accommodation'. Modelling one domain in another is analysed into three component activities by Pickering. First, a bridgehead is constructed in the second domain. Next, one transcribes 'established moves from the old system into the new space fixed by the bridgehead' (1997: 42). What is available for transcription is very rarely sufficient to complete the process, however, and so some filling is required. For Pickering, bridging and filling, the first and third of the components, are free moves, while transcription is based on forced moves dictated by the prior discipline. As is well known, the choices Hamilton made of requiring the retention of certain arithmetic and geometric features of the complex numbers meant that no three-dimensional analogue was available. Instead, he was led to define a third imaginary unit as part of the four-dimensional quaternion system.

This picture is, I believe, more or less correct, but we may put into question the stark dichotomy between the freedom of bridging and filling and the forcedness of transcription. After all, the capacity to accommodate to resistance during bridging or filling is often subject to fairly stringent requirements. Indeed, the very act of constructing a bridgehead may be guided by the experience of previous analogous analogies, and filling, in particular, must accord with existing conventions. In other words, we need to remember that the making of analogies does not occur in a vacuum, but rather in an intricate setting. Ulam's famous notion of mathematicians perceiving 'analogies between analogies' alludes to this intricacy, familiarity with which can only increase the feeling that one's choices are restricted.

It is probably this general sense of forcedness that gave Michael Crowe the idea that there was something irresistible about the invention of the quaternions once Hamilton had set off to look for higher-dimensional numbers. Crowe uses this case as an illustration of the first of his ten 'laws' concerning patterns of change in mathematics – 'New mathematical concepts frequently come forth not at the bidding, but against the efforts, at times strenuous efforts, of the mathematicians who create them.' (Crowe 1975: 16). At issue here is the degree of freedom we ascribe to Hamilton. After the event we might say that there are no normed division algebras in dimension 3 – inherent structure prevented him from succeeding in his initial goal. But Pickering might want to reply that Gibbs and Heaviside's solution of forming a vector calculus by hacking off a three-dimensional part of the quaternion system shows that no single solution is forced. Admittedly, associativity fails for the vector cross product, but then the octonions, the eight-dimensional normed division algebra devised by Hamilton's friend Graves, suffers from the same supposed defect. Where the vector calculus is a more radical departure, however, is in allowing zero divisors, e.g., for any x, $x \times x = 0$.

The other side of the coin is that we should be careful not to take transcription as a totally automatic process. We have seen in chapter 2 that Melis's attempt to automate even a simple piece of transcription runs into problems. Recall that for the analogical reasoning to work, we had to make sure that we had gone far enough along the sequence of nested rectangles to ensure that both dimensions were small enough. One would imagine that most trained mathematicians would cope very easily with this problem and approach it in broadly the same way. This suggests they may be working in a tight disciplinary framework yet still have a modicum of freedom. An alternative would be to count this kind of adjustment as *filling*, but then the term seems to be doing a lot of work, and *transcription* not enough.

Analogies vary as to the distance between the analogue domains. All things being equal, the larger the distance the greater the chance for the target domain's resistance. The pay-off is that analysis of resistance often leads to new ideas. Take, for example, a further use of the analogy between finite and infinite series, this time initially unsuccessful. Consider the series $1 - 1/2 + 1/3 - 1/4 + \cdots$ and double each term. The new series can then be rearranged and terms added to give back the original undoubled series, implying that the sum of the series is equal to twice itself and so must equal 0, while we know for other reasons that the series converges to ln 2:

$$2(1 - 1/2 + 1/3 - 1/4 + 1/5 - \cdots)$$
$$= (2 - 2/2) - 2/4 + (2/3 - 2/6) - 2/8 + (2/5 - 2/10) - \cdots$$
$$= 1 - 1/2 + 1/3 - 1/4 + 1/5 - \cdots$$

What has gone wrong? In a finite series the sum is the same whatever the order of the terms, yet analysis of this case shows that this property fails to hold generally for infinite series. The complex schema 'rearrange terms of a series to find sum' has been applied to a wider domain with only partial success. We can, however, find a condition which holds trivially for finite series and which will determine when the rearrangement of terms in the infinite case is valid. This is the condition of absolute convergence, which a series satisfies if the series formed by the absolute values of its terms converges. Such a condition would not occur to anyone working solely with finite sums.

This example may be construed as involving no real change to the source domain, but we must be careful not to assume that accommodation takes place only in the target domain. Sometimes it is better to say that the prior discipline has been reordered. The shift from the finite to the infinite frequently requires that modifications of this kind first be made. A good example of the need for reinterpretation before generalisation occurs in the extension of self-adjoint maps on finite vector spaces to infinite dimensional Hilbert spaces. Consider the following integral equation:

$$v(x) = \int_a^b K(x, s) u(s) dx.$$

For suitable $K(x, s)$ this defines a self-adjoint mapping taking $u(x)$ to $v(x)$, from the space of Lebesgue square summable functions on the interval $[a, b]$ to itself. Any function which is mapped to a scalar multiple of itself is then an eigenfunction of the mapping. The space of functions itself may be thought of as an infinite dimensional space with an inner product and the mapping as a linear function acting on it.

It makes sense to consider these mappings as infinite dimensional analogues of finite dimensional self-adjoint matrices and one then tries to push the analogy by diagonalising them. Any finite self-adjoint matrix can be so diagonalised such that its eigenvalues, which are all real, appear along the diagonal. The attempt to transfer this directly to the infinite case does not work, as many operators have o as their only eigenvalues and yet could not be represented by a zero infinite matrix. So first of all the finite dimensional theory must be recast, in what 'may seem a perverse manner' (Mackey 1992: 133), in terms of *projection valued measures*. To each subset of the reals, E, assign an operator P_E which maps an eigenvector to itself, if the corresponding eigenvalue belongs to E, and to o otherwise. It is only after this step that the idea of diagonalisation can be generalised to the infinite dimensional case to form the spectral theorem of Hilbert spaces.

One might wonder whether this modification of a simpler theory to allow for its generalisation, taking place as it does only after the discovery of the larger theory, merely acts as a means of justifying interest in the latter to the community. However, Hilbert believed that in many situations, finding the 'right' way to represent a domain would provide the key to cracking a more extensive one, thereby placing the original theory in its proper setting. A piece of heuristical advice from him tells us that:

In perhaps most cases when we fail to answer a question, the failure is caused by unsolved or insufficiently solved simpler and easier problems. Thus all depends on finding the easier problem and solving it with tools that are as perfect as possible and with notions that are capable of generalization. (Quoted in Booss and Bleecker 1985: 218)

We shall see in section 4.3 a case where the two domains of an analogy are very clearly revised because of the construction of a bridge between them.

Now we turn to analogies with greater apparent distance between source and target domains. Emily Grosholz has devoted a number of papers to the topic of structure similarity between two apparently distinct branches, focusing on Descartes' introduction of analytic geometry and Leibniz's extension of this to the study of dynamics. Here I shall deal only with the one entitled 'Two Episodes in the Unification of Logic and Topology' (Grosholz 1985). It might appear that this is yet another case of the phenomenon where exponents of the philosophy of real mathematics, although claiming not to afford privileges to mathematical logic over other branches of mathematics, still persist in drawing their case studies from this discipline. Grosholz, however, has as a reason for wanting to consider examples where an area of logic forms one side of the analogy, precisely the concern of deflating the special importance afforded to logic. By showing that areas of mathematical logic can partake in a structural partnership with topological concepts, Grosholz hopes to reduce what she sees as the unfair prominence given by philosophers to the former over the latter as an example of a mainstream branch.

The main thrust of her article is to illustrate her thesis that an important dynamic for mathematical development arises when between two branches there appears an unforeseen partial structural analogy. Here I shall talk principally about her first example, Stone duality. Grosholz's account is a little brief in certain respects, so I shall extend her account by bringing out the core of the analogy that drove Marshall Stone's programme. The roots of this theory go back to Boole's interpretation of classes or propositions in terms of algebra. A Boolean algebra is generated from a propositional language (collection of classes) by interpreting conjunction (intersection) as multiplication and disjunction (union) as addition. Lindenbaum and Tarski showed that necessary and sufficient conditions for a Boolean algebra to be representable as the algebra of all subsets of some set are that the algebra is complete and atomic, that is, it has to possess infinite meets and joins, and each element of the algebra has to be greater than or equal to an element not expressible as the join of strictly smaller elements. Their result, however, still left unsolved the problem of finding a general representation theorem for Boolean algebras.

While Grosholz describes Stone as an algebraist, his early work was in fact in functional analysis and this had a direct bearing on his later ideas. It was his work on algebras of commuting projections in a Hilbert space, which can be seen as Boolean algebras, that led Stone to seek a general representation theory. Moreover, through this background of functional analysis he had been made aware of the tools and concepts of general topology. In 1936–7, Stone published two long papers in which he solved the representation problem by making novel usage of ideas from topology. His first step was to notice that a Boolean algebra could be seen as a certain kind of ring (see appendix), where each of its elements satisfies $a = a^2$. As Peter Johnstone points out, four years earlier Stone had in fact used an 'informal analogy with ring theory, and it was not until 1935 that he realized the connection could be made formal' (Johnstone 1982: xv). This, now formal, analogy prompted him to investigate the role of ideals in a Boolean algebra. He then observed that the set of *prime* ideals of such an algebra, that is, those which cannot be factorised, could be given a topology[4] in which the clopen (closed and open) sets correspond to *principal* ideals. These are ideals generated by a single element of the algebra, and so provided him with a representation of the original algebra. The forging of this analogy with ring theory allowed him to create a representation carried not by some set of elements of the Boolean algebra as in the Lindenbaum–Tarski representation, but by the set of prime ideals.

In this example we find at least two creative acts: seeing a Boolean algebra as a ring; and, seeing the set of ideals as a space. The act of topologising such a seemingly non-spatial set helped to free applications of topology from their then close association with ordinary geometric spaces. Thereafter, Stone was wont to repeat the maxim 'one should always topologize'.

It is worth mentioning also that Stone duality is very well suited to the language of category theory. Indeed, the theorem can be restated to say that there is an adjoint equivalence between the category of Boolean rings and

---

[4] The open sets are in one-to-one correspondence with the general ideals of the Boolean ring. To such an ideal corresponds the set of those prime ideals which do not contain it.

the opposite of the category of zero-dimensional compact Hausdorff spaces. This equivalence can in turn be seen as being generated by an adjunction between larger categories.[5] Further evidence of the 'naturalness' of using category theory to explain Stone duality comes from the fact that in a standard introductory textbook to mathematical logic (Bell and Machover 1977), the only point at which the authors feel obliged to reformulate a result in category theoretic terms is precisely when they are dealing with Stone duality.

Johnstone (1982: xvi) believes that 'Stone's theorem was undoubtedly one of the major influences which prepared the mathematical world for the introduction of categories by Eilenberg and Mac Lane'. It was perhaps to be expected then that Mac Lane would, as we have seen, give a prominent place to analogy. Now, categories have the feeling in many branches of mathematics as always-having-been-there, so let us return to one of the most significant cases of analogy making in the history of mathematics, one which inspired Stone's work and which marked a dramatic escalation in the search for interbranch structural similarity.

4.3 THE ANALOGY BETWEEN NUMBER AND FUNCTION

The case study we shall investigate in this section is believed to be so good an example of the role of analogy in mathematics that when mathematicians are searching for an illustration of this phenomenon they very often turn to it. It was the obvious choice for one of the chief players in its later elaboration, the French mathematician André Weil, when, incarcerated in prison after forcible repatriation in 1940 for having tried to avoid military service by fleeing to Finland where he was lucky to escape being shot as a spy, he had the leisure to try to explain to his sister, the philosopher Simone Weil, the purpose of his mathematical research. This letter is published in his Collected Works (Weil 1940) and the subject matter taken up again in his 'De la Métaphysique aux Mathématiques' (Weil 1960).

The use of the word 'métaphysique' here refers to expressions such as 'the metaphysics of the calculus' used by mathematicians of the eighteenth century to indicate their incomplete understanding of the calculus or 'the metaphysics of equations' used by Lagrange to describe the, then unclear, ideas later made explicit by Galois. The title of Weil's paper refers to the act of clarifying mathematicians' blurred visions, and a very important way

---

[5] These results along with their many generalisations and extensions into almost every branch of mathematics are described in Johnstone (1982).

of doing this, he claims, is by working out analogies between theories. According to Weil:

As every mathematician knows, nothing is more fruitful than these obscure analogies, these indistinct reflections of one theory into another, these furtive caresses, these inexplicable disagreements; also nothing gives the researcher greater pleasure. (Weil 1960: 408, my translation)

It is the sense of surprise which attends such discoveries that suggests to mathematicians that they are on to something important as Dieudonné, Weil's fellow *Bourbakiste*, remarks:

But there are also what I shall call major translations, which could almost be termed mutations, and which appear to a certain extent out of the blue. They give the impression of a creation that nothing at all had prepared one for. (Dieudonné 1975b: 48, my translation)

The analogy which concerns us here is by no means simple, but I hope I may be able convey to the reader some sense of how intricate it is. We start with the simple observation that polynomials in one unknown with, say, complex coefficients, share many of the properties of the integers. In each case, given any two elements, there is an algorithm to find their highest common factor; there is unique decomposition into irreducible prime elements; analogous to the rationals are the rational functions (quotients of polynomials), etc. This would not seem to take us very far, but by the time this analogy was first noticed in the second half of the nineteenth century, Riemann had already developed his theory of many-valued complex functions of one variable to a high level of sophistication taking the first important steps in algebraic topology along the way.

In order to deal with many-valued complex functions, for example, $w^2 = z(z-1)$ for which, apart from when $z = 0$ or 1, w takes two distinct values, Riemann devised the surfaces later known after him. If the value of z is restricted to being defined on a single copy of the complex plane, on completing a single circuit of the circle of radius $1/2$ centred at the origin, one does not arrive back at the same value of w, but rather at the other branch of the function. Riemann's solution was to construct the *Riemann surface* of the function by taking two copies of the complex plane and cutting along a line running from the point 0 to the point 1 in each of them. This prevents the function changing branch along any curve in either of the planes. Now the planes are pasted together, the lower edge of the cut in one plane with the upper edge of the cut in the other, producing a surface on which the function is single-valued:

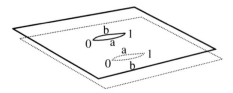

It is simpler for many purposes to add points at infinity to the two planes.
Then by deforming the resulting surface one can obtain a sphere.

Riemann then considered the set of meromorphic functions, that is,
differentiable except for isolated poles (infinities), defined on the Riemann
surface associated with a particular algebraic equation in two variables. He
noticed that these functions form a field (see appendix) and that this field
contains all one needs to know about the Riemann surface or algebraic
curve. This is an example of what Mac Lane has called a *shift of attention*
(1986: 433) and as we shall see has a close parallel to a similar construction
in number theory. This field is a finite extension of the field of rational
complex functions, $C(X)$.

During the 1870s Dedekind had studied algebraic numbers, i.e., solutions to equations in one unknown with rational coefficients. Given such
an equation, we adjoin its roots to the rationals, **Q**, thereby forming an
algebraic number field. If we consider for instance the equation $x^2 = -1$,
we then form the field **Q**(i). This field and its automorphisms (invertible
mappings preserving arithmetic relations) contain all we need to know
about the original equation. Dedekind noticed the parallel here between
this situation and Riemann's in that both involved finite extensions of fields
and this led him, in a paper he published along with Weber in 1882, to attempt the transfer of ideal theory, a recent construction of his in algebraic
number theory, to Riemann's algebraic curves so as to avoid the latter's
'transcendental' arguments.

To motivate Dedekind's introduction of ideals into number theory we
need to go a little further back to earlier work in number theory. Kummer's
principal interest in number theory was the continuation of the work of
Gauss on what is called quadratic reciprocity. During his investigations he
was led to the problem of whether unique factorisation holds among certain
collections of complex numbers, and found that it does not. It seems that
Kummer first explicitly mentioned this failure of unique factorisation in
1844, crediting Jacobi for the discovery.

A simple example of this phenomenon occurs in the ring of integers of
Q$(\sqrt{-5})$, where we find that $3.7 = (4 + \sqrt{-5})(4 - \sqrt{-5}) = 21$, and yet

it can easily be shown that none of the two pairs of factors of 21 can be factorised further. So although 3 cannot be factorised in $\mathbf{Q}(\sqrt{-5})$, it does not share the property possessed by ordinary primes that if p divides $a.b$ then it divides a or it divides b. The analogy between the ordinary integers and the ring of integers in a number field has failed and the two ideas, the former now called irreducibility, the latter called primeness, do not coincide. Kummer's task then was to try to rescue some notion of unique factorisation.

To do so Kummer had to in some sense 'factorise' these irreducibles further. He did so by providing a condition under which it might be said that an 'ideal' prime factor divided an integer. From there Kummer completed the task of showing that an algebraic integer could be factorised uniquely into prime ideal numbers. As with many constructions in mathematics, Kummer's ideal complex numbers engendered more than one successor. Kronecker's notion of a divisor is perhaps closer in spirit to its precursor, but here I shall speak only of Dedekind's introduction of *ideals*. I shall be looking at Kronecker and Dedekind's rival attempts to generalise Kummer's work in the context of research programmes in chapter 8.

The term *ideal* clearly derives from its progenitor. By calling his ideal numbers 'numbers' Kummer had taken an unusual step in the sense that any set of objects termed as such usually possesses an addition. This was not directly the case with Kummer's numbers, yet there was an unobvious sense in which it was so. Dedekind's ideals made this sense explicit. Very much in line with his philosophical viewpoint, Dedekind identified an ideal number with the set of integers it divided, where these integers are defined as roots of polynomials with integer coefficients and leading coefficient equal to 1. Dedekind and Kronecker converged on this latter definition analogising from a feature of the ordinary integers.

Dedekind then proved that in the integer ring of an algebraic number field the ideals possessed the property of unique factorisation, thus showing prime ideals to be the analogues of the prime numbers. In their paper of 1882, Dedekind and Weber now transferred these algebraic constructions to Riemann's complex function theory outlined above. Corresponding to $\mathbf{Z}$, $\mathbf{Q}$ and an algebraic extension K, we have $\mathbf{C}[X]$, $\mathbf{C}(X)$ and an algebraic extension $\mathbf{C}(X, Y)$, while corresponding to the algebraic integers of K we have the set of entire functions S, i.e., those which only have poles at infinity. But what then are the prime ideals of such an S? Well, for $\mathbf{C}[X]$ they are simply the principal ideals generated by $(X - a)$ with a in $\mathbf{C}$, and for S corresponding to a field extension $\mathbf{C}(X, Y)$ where the variables are related by an equation such as above $Y^2 = X(X - 1)$, the prime ideals are those of the form $(X - a, Y - b)$ where a and b are in $\mathbf{C}$ and satisfy $b^2 = a(a - 1)$. The essential point is that there is a one-to-one correspondence between the prime ideals of S and the points of the corresponding Riemann surface. Thus, the Riemann surface has been described purely in algebraic terms and this allows for the transfer of all the associated algebraic techniques into what might have been thought a geometric theory, in particular, it allows for a purely algebraic proof of the Riemann–Roch Theorem, a result which appeared to belong firmly to the subject of analytic varieties.

This idea of identifying points on a Riemann surface with prime ideals allows for the transfer of the idea of *ramification*. A Riemann surface is ramified with e sheets at a point p lying over $X = a$ if a closed path in a small neighbourhood of the point winds around a minimum of e times. The sum of the ramification indices of the points lying above a is equal to the degree of the defining polynomial seen as a polynomial in Y with coefficients in $\mathbf{C}(X)$. In our example $Y^2 = X(X - 1)$, $(0, 0)$ and $(1, 0)$ are ramification points of index 2 lying over 0 and 1, respectively, and hence there are no Taylor expansions of Y in powers of X or $(X - 1)$. Over a general point in the complex plane the Riemann surface will be unramified, e.g., over $X = 1/2$ there are two function elements represented by $Y = \pm i/2\{1 - 2(X - 1/2)^2 + \cdots\}$. Translating this into the language of ideals, the principal ideal $(X - 1/2)$ is prime in $\mathbf{C}[X]$, but in the set of entire functions, $\mathbf{C}[X, Y]/(Y^2 - X^2 + X)$, it splits into distinct primes as $(X - 1/2) = (X - 1/2, Y - i/2)(X - 1/2, Y + i/2)$. In the case of (X) however, we find it ramifies as $(X) = (X, Y)^2$.

Returning now to algebraic number fields, the question arises of how a prime ideal factorises in a field extension of $\mathbf{Q}$. Given (p) the prime ideal of $\mathbf{Z}$ generated by p, we know it must factorise completely into prime ideals of the ring of integers, O, of the number field, $(p) = P_1^{e_1} P_2^{e_2} \ldots P_g^{e_g}$, where any prime ideal P of O is such that O/P is a finite field of characteristic p, and so we can assign an f to each P where $O/P = (\mathbf{Z}/p\,\mathbf{Z})^f$.[6] Having done this we find the analogous formula $\Sigma_i e_i f_i = n$, where n is the degree of the field extension. If we now take the extension to be Galois then the e_i are all equal as are the f_i. It turns out that in any proper extension of $\mathbf{Q}$ there must exist a prime which ramifies, just as only the Riemann surfaces of rational functions are non-ramified.

Continuing with a further transfer of ideas in the same direction. Riemann's approach to complex function theory was not allowed undisputed control of the field. In Berlin, Weierstrass was pushing through a

---

[6] In function fields $\mathbf{C}(X, Y)$ we always have $f = 1$ as $\mathbf{C}$ is complete. This *inertia* in the number theoretic case is thus a disanalogy and was a spur to develop an intermediate language, as I shall describe below.

programme of arithmetisation based on the use of power series. Whereas Riemann studied complex functions from a global perspective, Weierstrass relied on a local approach in terms of power series expansions about a point. Any given rational complex function can be expanded in the neighbourhood of a point x_0 to give a series $\Sigma a_n(x-x_0)^n$, where the series runs from some (possibly negative) integer. However, not all expansions of this form correspond to rational functions. In 1897, Hensel, a student of Kronecker, pursued the analogy which identified points on an algebraic curve with prime ideals and was led to consider a similar extension to an algebraic number field. It will be easier to illustrate this construction if we choose this field to be $\mathbf{Q}$, for which the prime ideals are the principal ideals generated by the primes.

Just as $1/(1 - x)$ is expanded in the neighbourhood of 0 as $1 + x + x^2 + \cdots$, Hensel considered $1/(1 - p)$ as $1 + p + p^2 + \cdots$ Recalling that any member of $\mathbf{Q}$ can be uniquely expressed as $p^n a/b$ with n an integer, and $a, b \neq 0 \pmod{p}$, this can be given the following heuristic justification. If given two rational numbers we can divide their difference by arbitrarily high powers of p, then it is clear that the two numbers must be equal. Similarly, if we take the difference $1/(1 - p) - (1 + p + p^2 + \cdots + p^n)$, we can see that it is divisible by $p^{(n+1)}$, suggesting that it is 'small' from the perspective of p. For example, if $p = 5$,

$$1/(1 - 5) - (1 + 5 + 25 + 125) = -625/4 \text{ is divisible by } 5^4 = 625.$$

Now if we consider the set of all formal sums $\Sigma a_n p^n$, where $0 \leq a_n \leq p - 1$ and the sum begins with n possibly negative, and denote it $\mathbf{Q}_p$, it is easy to see that we can carry out on it all the usual arithmetic operations: addition, subtraction, multiplication and division. *p*-adic integers are those sums for which $a_n = 0$ for negative n.

We have a field, termed by Hensel the *p-adic* field, in which the ordinary rationals turn out to be those expansions which recur after a certain point. This suggests an analogy with the reals seen as the completion of the rationals under the usual metric $|a - b|$. We define a *valuation* $|x|_p = p^{-n}$, where n is determined by $x = p^n a/b$. $\mathbf{Q}_p$ is then the completion of $\mathbf{Q}$ under the metric $d(x, y) = |x - y|_p$. In $\mathbf{Q}_5$ for instance, we have shown above that $-1/4$ and 156 are 'close' to each other.

Weil later observed that this notion of valuation which links the standard distance function in $\mathbf{R}$ or $\mathbf{C}$ with the *p*-adic distance removes a noticeable case of disanalogy between algebraic number fields and algebraic number fields. Consider once again $\mathbf{Q}$ and $\mathbf{C}(X)$. For elements in each field we have the factorisations $p_1 \cdots p_m/q_1 \cdots q_n$ and $a(x - \alpha_1) \cdots (x - \alpha_m)/(x - \beta_1) \cdots (x - \beta_n)$, but in the latter case the point at infinity is treated on an equal footing with the other points, there being a zero (or pole) at infinity of order $|n - m|$ when $n > m$ (or $n < m$). There is no immediate analogue for this in the field of rational numbers. However, we may rethink this case in terms of valuations. For each α, there is a function which for any rational function takes as its value the power of $(x - \alpha)$, and for ∞ takes the value $(n - m)$. The sum of these functions for any rational function is then 0. Similarly, taking the valuation 'at infinity' to be the ordinary real modulus, the product of the valuations on a rational number is 1. To give another sense of this, if we think of a decimal expansion of a real number, it is similar to a p-adic series except in that moving along the series the exponent of 10 decreases instead of increasing, rather like the expansion of a rational function at ∞ in terms of $1/x$.

By the mid-twentieth century, mathematicians such as Weil were milking the analogy for all its worth. To give a flavour of this highly complex work, note that a meromorphic function may be considered not merely as being expandable at one point, but at all points. In the simplest case, this corresponds to embedding $\mathbf{C}(X)$ in the set of formal series in $(x - x_0)$ for each x_0. Similarly we can embed an algebraic number field into its p-adic completions and do so simultaneously. In other words, we have a map into the product of these completions, or rather into a subset of this product, namely the *adèle* whose members are choices of elements from each completion almost all of which belong to the respective completion of the integers. Adèles may be given a topology which turn them into a locally compact group, thereby allowing all the techniques of harmonic analysis (a generalisation of Fourier analysis) to be imported.

4.4 A WATERSHED IN THE USE OF ANALOGY

If section 4.3 was rather hard going for you, I hope you still managed to gain the sense that the Dedekind–Weber paper of 1882 laid the foundations for an extraordinarily rich transfer of concepts between the fields of algebraic number theory and algebraic function theory. Dieudonné notes how in this paper we find:

one of the key ideas of modern mathematics, which consists in performing calculations on objects which are not at all like numbers or functions. Moreover, this article by Dedekind and Weber drew attention for the first time to a striking relationship between two mathematical domains up until then considered very remote from each other, the first manifestation of what was to become a 'leitmotif' of later work: the search for common structures hidden under at times extremely disparate appearances. (Dieudonné 1969: 375, my translation)

This is a thoroughly *Bourbakiste* vision of mathematics, passed down to them from Dedekind via Hilbert, Noether and Artin.

Of course, if through abstraction the point is ever reached where the concepts common to two or more fields are made sufficiently explicit to allow the straightforward transcription of results between these fields, the analogy has served its purpose and is at the end of its useful life; the fields will have been grounded in a common discipline:

The day dawns when the illusion vanishes; intuition turns to certitude; the twin theories reveal their common source before disappearing; as the *Gita* teaches us, knowledge and indifference are attained at the same moment. Metaphysics has become mathematics, ready to form the material for a treatise whose icy beauty no longer has the power to move us. (Weil 1960: 408, my translation)

Bourbaki have often been the butt of criticism for having written such a treatise, but its members were well aware that it is often the case that the fields bearing the analogy present considerable resistance to their common grounding. This appears to be the case in our example where Serge Lang, another *Bourbakiste*, justifies his separate treatment of algebraic number fields by pointing out that:

certain aspects of number fields are still shrouded in mystery while the corresponding aspects of the function field case are cleared up. Thus a certain emphasis on the peculiarities of number fields is not out of place. (Lang 1970: 175)

A similar point is also noted by Emily Grosholz in the paper mentioned in section 4.2, in particular in the case of Stone duality:

Structural analogies are interesting not only in their development, but also in their limitations. In the end, only a restricted portion of topology was amenable to correlation with propositional logic; both fields, in resisting each other, show their own characteristic and irreducible texture. (Grosholz 1985: 151)

We should not forget, however, that the incompleteness of an analogy at a given moment is no indication that it may not be extended or subsumed at a later date. But even if all is eventually explained, these glimpses of some deeper explanation for structure similarity do not just operate briefly. They may hang tantalisingly in the air for decades and play a crucial role in directing research.

In the case considered in section 4.3, although the analogy between algebraic number fields and algebraic function fields had proved immensely rich, there was still the perception that much of the intricate structure of the complex function fields was unavailable for transfer for want of a broad enough bridge between the Riemannian and number field domains.

To provide a broader conduit it was considered necessary to construct an intermediate platform to provide support for a two-arched bridge. Just such a support was found in the shape of function fields over finite constant or Galois fields.

Weil (1960: 411–122) likens this new situation to that faced by Champollion as he commenced his Herculean task of deciphering the Rosetta Stone. The problem in this case is also to translate passages in the text of one of the three languages into the other two languages. For example, in the early 1920s Emil Artin in his doctoral thesis translated the statement of the Riemann hypothesis into the middle language of 'Galoisian', where it was proved some time later. Owing to our inability to translate this proof back into Riemannian, the original hypothesis remains as yet unproven. Artin also managed a two-step translation of a further construction from Riemannian to number theory passing through Galoisian.

4.5 CONCLUSION

I want to underscore the historical claim that the work of Weber and Dedekind represented a significant moment in the movement of mathematics towards what is often called a 'structural' outlook. The thoroughness of their analogising gave great heart to this movement whose staging posts include Hilbert, Noether and later *Bourbaki*. Not everyone has been so enchanted as Weil by this style of mathematics, but it has certainly played a major part in shaping the mathematics of the twentieth century.

For Weil (1940) the process of analogy formation resembles the work of sculpting a hard piece of rock whose structure dictates the emerging shape. But this inherent-structurist picture may of course be challenged. Do function fields and number fields possess unusually strong structural similarities to allow fruitful analogies to take place? Was the analogy forced upon them? Is it perhaps easy to do this sort of thing for any two types of mathematical entity? Is it just the way mathematicians expand into unknown territory owing to their psychological make-up? The evidence we need to argue for these positions lies all around us. For instance, the programme described in Kreimer (2000) includes an attempt to restore prime factorisation to Feynman diagrams based on an analogy with number fields. Let's get to work on these questions.

Finally, I would like to relate what we have discovered in this chapter on human mathematics to the work of chapters 2 and 3 on automated mathematics. Note how we have seen mathematicians operating with 'bite-size' chunks of mathematical concepts: factorisation, primality, ramification, inertia, completions, embeddings. This is the order mathematicians work at, and it seems to suit them because they need only employ, in a recursive fashion, a manageable number of them. The gulf between human and computer assisted analogical reasoning is broad. Mathematicians are operating two levels up from Wos's analogy or resonance via syntactical similarity and one level up from Mellis's analogy via proof plans. For a computer to collaborate at the human level it would require one whose sophistication resembled that of the assistant appearing in Gowers's fantasy (2000b).

PART II

Plausibility, uncertainty and probability

To acquire insight into a discipline one needs to comprehend how its practitioners reason plausibly. This is no less true for mathematics than it is for science. Understanding how mathematicians choose which problems to work on, how they formulate conjectures and the strategies they adopt to tackle them all require considerations of plausibility. Furthermore, it is also the case that the plausibility of a scientific theory may depend on the plausibility of mathematical results. This has always been so, but now we live in an era where for some physical theories the *only* testable predictions are mathematical ones it is coming to the fore. Thus, if we are to understand how physicists reckon on the plausibility of their theories, this must involve paying due consideration to the effect of verifying uncertain mathematical predictions.

Now, if one decides, as many have, to treat plausible and inductive reasoning in the sciences in Bayesian terms, it seems clear that one would want to do the same for mathematics. After all, it would appear a little extravagant to devise a second calculus. In any case, Bayesianism is usually presented by its proponents as capable of treating all forms of uncertain reasoning. If so, then we can say that Bayesianism in science requires Bayesianism in mathematics. Once this is accepted, we shall respond in one of two ways according to the discoveries made while examining Bayesianism in mathematics:

I. Bayesianism cannot be made to work for mathematics, therefore Bayesianism cannot give a complete picture of scientific inference.

II. Some forms of Bayesianism can be made to work for mathematics, therefore one of these must be adopted by Bayesian philosophers to give a more complete picture of scientific inference.

The arguments presented in chapter 5 indicate that the antecedent of I is false and the antecedent of II true, opening the prospect of an expanded, but modified, Bayesianism. Chapter 6 then investigates how the Bayesian should integrate plausibility considerations in mathematics and science, and claims more generally that philosophers of science need to take much greater account of the contribution of mathematical reasoning to science.

Those at the historically minded end of philosophy may view this turn to Bayesianism with a jaundiced eye. Back in 1962, Kuhn warned of the problems of probabilistic verification theories (Kuhn 1962: 145–6) and Lakatos was even less supportive (Lakatos 1968). We shall move on to Lakatos in part III, but in the meantime, if Bayesianism gets us talking about plausibility in mathematics and the role of mathematics in science, that can be no bad thing.

CHAPTER 5

Bayesianism in mathematics

Even in the field of tautology (i.e. of what is true or false by mere definition, independently of any contingent circumstances) we always find ourselves in a state of uncertainty. In fact, even a single verification of a tautological truth (for instance, of what is the seventh, or billionth, decimal place of π, or of what are the necessary or sufficient conditions for a given assertion) can turn out to be, at a given moment, to a greater or lesser extent accessible or affected with error, or to be just a doubtful memory.

(de Finetti 1974: 24)

After all, which is more compelling, a formal proof that in its full exposition requires hundreds of difficult pages of reasoning, fully understood by only two or three colleagues, or the numerical verification of a conjecture to 100,000 decimal digit accuracy, subsequently validated by numerous subsidiary computations?

(Bailey and Borwein 2001: 53)

5.1 INTRODUCTION

In his *Mathematics and Plausible Reasoning* (Pólya 1954a, 1954b), Pólya suggests that mathematics is the perfect domain in which to devise a theory of plausible reasoning. After all, where else do you find such unequivocal instances of facts satisfying general laws, and where else do you find general laws being established so conclusively? As a noted mathematician actively engaged in research, Pólya delightfully conveys inferential patterns by means of examples of his own use of plausible reasoning to generate likely conjectures and workable strategies for their proof. What concerns us in this chapter is the fact that in the second of these two volumes, he works his account of plausible reasoning into a probabilistic mould. It appears that in doing so he became indebted to Bruno de Finetti, one of the founders of modern subjective Bayesianism (Pólya 1941: 451).[1] Moreover, he can be

---

[1] 'I owe much to a conversation which I had the pleasure of having with Bruno de Finetti' (Pólya 1941: 451). In the same paper, he also indicates that he has read J. M. Keynes' *A Treatise on Probability* (*ibid.*: 462).

viewed as a pioneer who influenced some later prominent Bayesians. Edwin Jaynes certainly learned from his work, and it is clear that Judea Pearl has read him closely. So, while he did not name himself as such, we may reasonably view Pólya as a member, or at least an associate, of the Bayesian camp. But what are we to make of a Bayesian interpretation of mathematical reasoning?

In sections 5.1 and 5.2 we shall be inquiring as to which varieties of the many forms of Bayesianism are best able to accommodate mathematical reasoning. Bayesians commonly hold as tenets that logically equivalent sentences should be believed by an ideal agent with equal confidence and that any evidence should have an equal impact on their degrees of belief in such sentences. However, this ideal throws little or no light on plausible mathematical reasoning, since from its perspective mathematicians can only be deemed to be failing dismally as soon as they express the strength of their belief in mathematical propositions. We are thus led to search for a reasonable relaxation of the strictures of omniscience.

Arguing for such a relaxation is a thankless task. Half of your audience take it to be so obvious that they accuse you of insulting *their* intelligence; while the other half take it as such a hopelessly wrong-headed idea that they insult *your* intelligence. Even so, I feel there is something to be said, and so in section 5.2 I argue that if a Bayesian modelling of plausible reasoning in mathematics is to be of value, then the assumption of logical omniscience should be relaxed.

In Pólya's version of Bayesianism in mathematics we have only the right to specify the direction of change in the credence we give to a statement on acquiring new information, not the magnitude. However, Edwin Jaynes demonstrated that one of the central grounds for this decision on Pólya's part to avoid quantitative considerations was wrong. In section 5.3 I consider whether there is any scope for a quantitative form of Bayesianism in mathematics.

One criticism often made of Bayesian philosophy of science is that it does not help very much in anything beyond toy problems. While it can resolve simple issues, such as accounting for how observing a white tennis shoe provides no confirmation for the law 'all ravens are black', it provides no insight into real cases of theory appraisal and confirmation. Much rests on the assignment of priors which, it is claimed, can be chosen to make most pieces of scientific reasoning look reasonable. Recognising what is correct in this criticism, I think there is still useful work to be done. In section 5.4 I shall be looking in particular at: reasoning by analogy; choice of proof strategy (for automated theorem proving); and large-scale induction (particularly enumerative induction).

5.2 PROBABILITY THEORY AS LOGIC

Plausible reasoning in mathematics is, of course, necessary only because mathematics does not emerge in the minds of mathematicians as it appears on the pages of a journal article or textbook, that is, in its semi-rigorous deductive plumage. Indeed, it is owing to the failure of what is called 'logical omniscience', the capacity to know immediately the logical consequences of a set of hypotheses, that mathematicians are forced to resort to what amounts to a guided process of trial and error, not so dissimilar to that employed in the natural sciences. So here we have something of a paradox: plausible mathematical reasoning, the subject of Pólya's analysis, was an important source of ideas for some of the leading figures of Bayesianism, and yet it is necessitated by the fact that people involved in this most rigorous branch of knowledge cannot come close to adhering to the ideal constituted by a widely held tenet of Bayesianism, namely, that logically equivalent statements receive identical degrees of belief, or alternatively, that tautologies be believed with degree of belief set at 1.

Let us illustrate this with a simple example. Imagine you are to bet on the trillionth decimal place of π.[2] As I write this, computers have 'only' reached the 206 billionth place, so this wager concerns a result that nobody yet knows but which we may imagine will be settled in the not too distance future. Surely it is reasonable for you to prefer a bet on this digit being between 0 and 8, to one at the same odds on its being 9. If this is your preference, we can say that your degree of belief that this digit is 9 is less than 0.5. If, however, 9 is the correct digit, then it follows as a 'mere' calculation from one of the series expansions for π. That is, '$\pi = 4(1 - 1/3 + 1/5 - 1/7 + \cdots)$' and '$\pi = 4(1 - 1/3 + 1/5 - 1/7 + \cdots)$ & the trillionth decimal place of π is 9' would be logically equivalent and so to be believed with the same confidence, and so the second bet should be preferred. Indeed, any degree of belief for a digit of π which is strictly between 0 and 1 is incoherent from the omniscient perspective, or, to use the language of Howson (2000), adopting such degrees of belief cannot be done consistently.

Now, someone who wishes to retain the condition that tautologies be believed to degree 1 might respond, as does Howson, that this is not a

---

[2] Some Bayesians object to the use of the language of gambling. Whatever I say here in this language is readily translatable into other preferred idioms.

problem, that 'the possibility of inconsistency is like that of death, a condition of life, an omni-present hazard' (Howson 2000: 150), and that all
one need do is adjust Pr(A) to 1 on discovering that A is a logical truth. But
still one wants to reserve some kind of label to chastise the fool who offers
0.1 for a particularly digit being a '0' and 0.2 for each of the other values,
to distinguish him for someone who reasonably offers 0.1 for each value of
the digit, but who must be 'wrong'.

One of Howson's arguments for retaining omniscience rests on an analogy with deductive logic. There, if I believe I have found a deductive proof
of a proposition, I now have confidence in its truth. If later a clear counter-
example is shown to me, I simply alter my truth ascription. I do not argue
for changes to be made to the inference rules to accommodate inconsistent
truth ascriptions.

One response to this point is to claim, as did Pólya (1954b: 112–16), that
deductive logic and Bayesian inference are qualitatively different. A mathematician is cautious about what she alleges to have shown deductively. Were
she found to have erred on even 1 per cent of the theorems proved in her
papers, her credibility would be severely damaged. On the other hand, she
may use plausible inference to assess the plausibility of *any* statement based
on what she knows at present.[3] And as soon as she announces non-extreme
degrees of belief for a set of mathematical statements, our understanding
that she is not logically omniscient should make us expect them to be incoherent in the omniscient sense. This will be so even for mathematicians
with an excellent sense of what is likely to be true. The question is whether
there are reasonable ways in which we can arrive at, what from a God's eye
view are incoherent, degrees of belief on the basis of what we already know.

Howson (2000: 200–1) also introduces some interesting ideas about more
sophisticated statements such as 'Peano Arithmetic is consistent' (PAC).
Here we have a mathematical statement which is neither a logical truth
nor falsehood. Rather, it is a statement about which we have a considerable
amount of evidence (no discovery of inconsistency to date, proof of consistency in stronger, but potentially inconsistent system, etc.), bearing on our
assessment of a degree of belief. Of course, there arises the issue of how one
could establish PAC, or settle a bet in favour of its truth if you prefer that
language, but let's imagine that resolved. We would still want to allow as

---

[3] Jaynes (forthcoming ch. 10: 21) has a similar view on the difference between deductive logic and
probability theory as logic: 'Nothing in our past experience could have prepared us for this; it is a
situation without parallel in any other field. In other applications of mathematics, if we fail to use
all of the relevant data of a problem, the result will be that we are unable to get any answer at all.
But probability theory cannot have any such built-in safety device, because in principle, the theory
must be able to operate no matter what our incomplete information might be.'

reasonable the possibility of assessing our degree of belief in a statement of arithmetic below that given to PAC, when in fact the statement is provable in Peano arithmetic, differentiating this from some foolish choice.

Someone alive to this difference is Ian Hacking. In his article, 'Slightly More Realistic Personal Probability' (Hacking 1967), he sets out a hierarchy of strengths of Bayesianism. These strengths he correlates with ways of saying whether a statement can be possibly true. At the weaker end we find a position he terms 'realistic personalism', where non-zero probabilities will be attributed by a subject to any statement not known by them to be false, knowledge being taken in a very strict sense: 'a man can know how to use *modus ponens*, can know the rule is valid, can know p, and can know $p \supset q$, and yet not know q, simply because he has not thought of putting them together' (Hacking 1967: 319). At the stronger end we find logical omniscience and divine knowledge.

Now clearly the coherence provided by realistic personalism is not enough to equip you for a life as a gambler. For instance, it is advisable not to advertise on a mathematics electronic bulletin board the odds at which you would accept either side of a wager on a mathematical proposition whose truth status you have every reason to believe has been settled. Dutch Book arguments and their ilk, used to establish the irrationality of holding degrees of belief not satisfying the probability axioms, do not prove your irrationality on the grounds that someone may know more than you. If they do know more than you, you will tend to lose whether the subject of your bet is mathematics, physics or the date of the next general election.

An intermediate position between realistic personalism and logical omniscience involves considering mathematics as a body of knowledge, where the ideal would be to know the sum of what leading researchers know. This idealisation of being aware of all known mathematics is, unlike in the case of omniscience, which offers little more by way of advice than to be as perfect a logician as you can be, at least one which may be aspired to. And it does seem that an acknowledged trait of a good mathematician is that she can judge what is already known. For instance, the mathematician Stanislaw Ulam tells us that:

As for myself, I cannot claim that I know much of the technical material of mathematics. What I may have is the feeling for the gist, or maybe only the gist of the gist, in a number of fields. It is possible to have this knack for guessing or feeling what is likely to be new or already known, or else not known, in some branch of mathematics, where one does not know the details. I think I have the ability to a degree and can often tell whether a theorem is known, i.e., already proved, or is a new conjecture. (Quoted in Gardner 1983: 143)

But this capacity is becoming less important with the advent of new technology. The arrival of MathSciNet, an online database of abstracts, allows mathematicians to determine what is currently happening in specific research areas. As David Eisenbud remarks:

This allows me to search through the body of mathematical knowledge in a way that was never before possible. When I'm thinking about a new area, I can go to a computer and quickly see what has and what hasn't been done. (Eisenbud 2001: 651)

Thinking in terms of such communal knowledge points the intrepid Bayesian philosopher of mathematics to look into the intricacies of what actually happens in mathematics departments to find out how this body grows.

It is interesting to speculate why Pólya has been so little taken up on his Bayesianism in mathematics by philosophers. What is the underlying intuition behind the avoidance of a Bayesian treatment of plausible and inductive reasoning in mathematics? We can begin to understand what is at stake when we read Mary Hesse's claim that 'since mathematical theorems, unlike scientific laws, are matters of *proof*, it is not likely that our degree of belief in Goldbach's conjecture is happily explicated by probability functions' (1974: 191). There are two responses to this. First, while it is true that the nature of mathematics is characterised like no other discipline by its possession of deductive proof as a means of attaining the highest confidence in the trustworthiness of its results, proofs are never perfectly secure. Second, and more importantly, what gets overlooked here is the prevalence in mathematics of factors other than proof for changing degrees of belief.

The lack of attention plausible mathematical reasoning has received reflects the refusal of most English-language philosophers of mathematics to consider the way mathematical research is conducted and assessed. On the basis of this refusal, it is very easy then to persist in thinking of mathematics merely as a body of established truths. As classical deductive logic may be captured from a probability calculus which permits propositions to have probabilities either 0 or 1, the belief that mathematics is some kind of elaboration of logic and that the mathematical statements to be considered philosophically are those known to be right or wrong go hand in hand. We could say in fact that mathematics has suffered philosophically from its success at accumulating knowledge since this has deflected philosophers' attention from mathematics as it is being developed. But one has only to glance at one of the many survey articles in which mathematicians discuss the state of play in their field, to realise the vastness of what they know to be unknown but are very eager to know, and about which they may be thought to have degrees of belief equal neither to 0 nor to 1.[4]

We shall see in section 4.4 how mathematical evidence comes in very different shapes and sizes. But even remaining with 'proved' or well-established statements, although there would appear to be little scope for plausible reasoning, there are a number of ways that less than certain degrees of belief can be attributed to these results. David Hume described this lack of certainty well:

> There is no Algebraist nor mathematician so expert in his science, as to place entire confidence in his proof immediately on his discovery of it, or regard it as any thing, but a mere probability. Every time he runs over his proofs, his confidence encreases; but still more by the approbation of his friends; and is rais'd to its utmost perfection by the universal assent and applauses of the learned world. Now 'tis evident, that this gradual encrease of assurance is nothing but the addition of new probabilities, and is deriv'd from the constant union of causes and effects, according to past experience and observation. (Hume 1739: 180–1)

Perfect credibility may be difficult to achieve for proofs taking one of a number of non-standard forms, from humanly generated unsurveyable proofs to computer assisted proofs to probabilistic proofs. These latter include tests for the primality of a natural number, n. Owing to the fact that more than three-quarters of the numbers less than n are easily computed 'witnesses' to its being composite, if such is the case, a relatively small sample will provide us either with a proof of compositeness or else with very powerful evidence for n being prime, and we can prolong this sampling to reduce doubt as far as we care to.

While a certain amount of suspicion surrounds the latter type of 'proof', from the Bayesian perspective, one can claim that all evidence shares the property that it produces changes in some degrees of belief. The question of whether there is any qualitative difference in the epistemic import of different types of proof has been examined by Don Fallis (1997), who considers many possible ways of distinguishing epistemically between deductive proofs and probabilistic proofs and finds none of them adequate. Fallis centres his discussion around 'proofs' which involve clever ways of getting strands of DNA to react to model searches for paths through graphs, putting beyond reasonable doubt the existence or non-existence of such paths. Despite there being here a reliance on biochemical knowledge, Fallis

---

[4] I mean to exclude here the immense tracts of totally uninteresting statements expressible in the language of ZFC in which one will never care to have a degree of belief.

still sees no qualitative difference as regards the justificatory power of this type of proof. Confidence in mathematical statements is being determined by natural scientific theory. This appears less surprising when you consider how complicated, yet well modelled, configurations of silicon can be used to generate evidence for mathematical propositions.

The less controversial part of the position Fallis is examining may be expressed in Bayesian terms as follows.[5] The reliability of a mathematical statement is dependent solely on your rational degree of belief in that statement conditioned on all the relevant evidence. Whatever level you set yourself (0.99 or 0.99999), the type of evidence which has led you there is irrelevant. You may judge that a 10,000-page proof provides as much support as a probabilistic proof or the non-appearance of a counter-example. To contemplate the reliability of a result in a particular field we should think of someone from outside the field asking a specialist for their advice. If the trustworthy expert says she is very certain that the result may be relied upon, does it matter to the enquirer how the specialist's confidence arises? This depiction could be taken as part of a larger Bayesian picture. The very strong evidence we glorify with the name 'proof' is just as much a piece of evidence as is a verification of a single consequence. Bayesianism treats in a uniform manner not just the very strong evidence that Fallis considers, but all varieties of partial evidence. Let us now see what we are to make of this partial evidence.

5.3 QUANTITATIVE BAYESIANISM

As I have said, Pólya (1954b: 112–16) understood plausible inference to be quite different from deductive logic. In his opinion, deductive logic is:
(a) Impersonal – independent of the reasoner
(b) Universal – independent of the subject matter
(c) Self-sufficient – nothing beyond the premises is needed
(d) Definitive – the premises may be discarded at the end of the argument.
On the other hand, plausible inference is characterised by the following properties:
(a) The direction of change in credibility is impersonal, but the strength may be personal
(b) It can be applied universally, but domain knowledge becomes important for the strength of change, so there are practical limitations

---

[5] He points out (private communication), however, that he is not necessarily committed to a Bayesian analysis of this position, nor does he deny other epistemic virtues of traditional proofs.

(c) New information may have a bearing on a plausible inference, causing one to revise it

(d) The work of plausible inference is never finished as one cannot predict what new relevant information may arise.

In this section we shall be concerned with numerical and even algorithmic kinds of Bayesianism. It is clear that the strength of a mathematician's belief in the correctness of a result has an impact on their practice: Andrew Wiles would hardly have devoted seven years to Fermat's Last Theorem had he not had great faith in its veracity. No doubt we could give a complicated Bayesian reconstruction of his decision to do so in terms of the utility of success, the expected utility of lemmas derived in a failed attempt, and so on. For a simpler example, let us give a Bayesian reconstruction of the following decision of the French Academy:

The impossibility of squaring the circle was shown in 1885, but before that date all geometers considered this impossibility as so 'probable' that the Académie des Sciences rejected without examination the, alas!, too numerous memoirs on this subject that a few unhappy madmen sent in every year. Was the Académie wrong? Evidently not, and it knew perfectly well that by acting in this manner it did not run the least risk of stifling a discovery of moment. The Académie could not have proved that it was right, but it knew well that its instincts did not deceive it. If you had asked the Academicians, they would have answered: 'We have compared the probability that an unknown scientist should have found out what has been vainly sought for so long, with the probability that there is one madman the more on earth, and the latter has appeared to us the greater.' (Poincaré 1905: 191–2)

These alternatives, being mad and being right, were hardly exhaustive. Leaving aside the person's sanity we can contrast the probability that their proof is correct with the probability that it is incorrect:

```
Pr(proof correct | author unknown)=
Pr(proof correct | author unknown, result true)·
Pr(result true | author unknown) +
Pr(proof correct | author unknown, result false)·
Pr(result false | author unknown).
```

Assuming we are dealing with a consistent system in which a false result cannot be proved correctly, and assuming that the existence of the unknown author has no bearing on the truth of the result, we have then:

```
Pr(proof correct | author unknown)=
Pr(proof correct | author unknown, result true)·
Pr(result true).
```

Substituting reasonable estimates of the Académie's degrees of belief will lead to a very small value for this last expression because its two factors are small. So small indeed that even though the gain of discovering a correct proof greatly outweighs the effort wasted in discovering a flaw in a submitted proof, greater expected utility arises from consigning it to the bin. On the other hand, a submitted proof of the possibility of squaring the circle by a known mathematician, or a submitted proof of its impossibility by an unknown author would presumably have been dealt with more tolerantly. Of course, the credibility of even the best-known mathematicians will not be judged to be uniform. Listen to Jacobi in 1846: 'When Gauss says he has proved something, I think it very likely; when Cauchy says it, it is a fifty-fifty bet; when Dirichlet says it, it is *certain*' (Laugwitz 1999: 63). Standards have evidently improved.

Notice that this reconstruction would not seem to require one to go beyond vague talk of very high or very low probabilities. By contrast, when it comes to offering a betting ratio for the trillionth decimal digit of π being 9, it would seem to be eminently reasonable to propose precisely 1/10. What appears to determine this value is some form of the principle of indifference based on our background knowledge. With a simple grasp of the idea of a decimal expansion we simply have no reason to believe any single digit more likely than any other. Those who know a little more may have heard that to date there is neither statistical evidence nor theoretical explanation for any lack of uniformity in the known portion of the expansion, probably rendering them much less likely to be swayed in their betting ratio by a spate of 9s occurring shortly before the trillionth place. Those who have read (Bailey and Crandall 2001) may have been persuaded by their use of dynamical systems theory to argue for the normality of π, and be extremely resistant to changing their betting ratio. So, unless some dramatic piece of theoretical or empirical evidence is found, it seems that most mathematicians would stick with the same betting ratio until the point when they hear that computers have calculated the trillionth place.

Do we require a quantitative, or even algorithmic, form of Bayesianism to allow us to explicate plausible mathematical reasoning, or, like Pólya, should we make do with a qualitative form of it? First, it will be helpful for us to contrast Pólya's position with that of Jaynes. For Jaynes, Pólya was an inspiration. Indeed, he

was the original source of many of the ideas underlying the present work. We show how Pólya's principles may be made quantitative, with resulting useful applications. (Jaynes forthcoming, ch. 1: 3)

Jaynes was positioned at the objectivist end of the Bayesian spectrum. In other words, his aim was to establish principles (maximum entropy, transformation groups, etc.) applicable in as many situations as possible, in which a reasonable being could rationally decide on their prior probabilities. Pólya, on the other hand, reckoned that one would have to stay with a qualitative treatment (e.g. if *A* is analogous to *B* and *B* becomes more likely, then *A* becomes somewhat more likely), in that the direction of changes to confidence might be determined but not their strength. But Jaynes claimed that this decision was based on a faulty calculation made by Pólya when he was considering the support provided to Newton's theory of gravitation by its prediction of the existence and location of a new planet, now called Neptune. The incorrect calculation occurred when Pólya was discussing the boost to confidence in Newtonian gravitation brought about by the observation of a previously unknown planet precisely where calculations predicted it to be, based on observed deviations in Uranus's orbit.

Pólya takes Bayes theorem in the form,

```
Pr(Newt. Grav.| Neptune) = Pr(Newt. Grav.)·
Pr(Neptune | Newt. Grav.)/Pr(Neptune),
```

where `Pr(Neptune)` corresponds to a scientist's degree of belief that the proposed planet lies in the predicted direction. For the purposes of the calculation, he estimates `Pr(Neptune)` in two ways. First, he calculates the probability of a point lying within one degree of solid angle of the predicted direction, and arrives at a figure of $0.00007615 \approx 1/13100$. Second, on the grounds that the new planet might have been expected to lie on the ecliptic, he uses the probability of a point on a circle lying within one degree of the specified position, yielding a value for `Pr(Neptune)` of $1/180$. He then argues that `Pr(Newtonian Gravitation)` must be less than `Pr(Neptune)`, otherwise Bayes's theorem will lead to a posterior probability greater then 1, but that it is unreasonable to imagine a scientist's degree of belief being less than even the larger figure of $1/180$, since Newtonian Gravitation was already well confirmed by that point. He concludes, 'We may be tempted to regard this as a refutation of the proposed inequality' (1954b: 132), and suggests we return to a safer qualitative treatment.

However, as Jaynes points out, Pólya's calculations were in fact of the prior to posterior odds ratio of two theories: on the one hand, Newtonian gravitation and, on the other, a theory which predicts merely that there be another planet, firstly anywhere and secondly on the ecliptic. Indeed, from the confirmation, Newtonian gravitation is receiving a boost of 13100 or 180 *relative* to the theory that there is one more planet somewhere. Pólya had forgotten that if Pr(Newtonian Gravitation) is already high then so too would be Pr(Neptune).

We are told by Jaynes that Pólya realised his mistake and went on to participate vigorously in the former's lectures at Stanford University in the 1950s. However, Pólya had given several further arguments against quantitative plausible reasoning, so even if Jaynes could block this particular argument, one would need to confront the others. Reading through them, however, one notes that Pólya is making fairly standard points – the incomparability of evidence and conjectures, problems with the principle of indifference, etc. – which Bayesians such as Jaynes have responded to elsewhere.

Could it be that your background predisposes you to adopt a certain type of Bayesianism? The physicist relies on sophisticated symmetry considerations pertaining to the physical systems and instruments producing the data, the philosopher of science on vaguer considerations of theory evaluation, while the economist must integrate a mass of data with her qualitative, quasi-causal understanding of the economy. Are disputes among Bayesians like the blind men feeling different parts of an elephant?

Bayesianism applied to reasoning in the natural sciences appears to fall into two rather distinct categories:

(1) Analysis of data from, say, nuclear magnetic resonance experiments or astrophysical observations, permitting model comparison and parameter estimation

(2) Plausible reasoning of scientists by philosophers of science (e.g. Franklin 1986).

We may wonder how strong the relation is between them. Rosenkrantz (1977) attempted a unified treatment, and he indicates by his subtitle *Towards a Bayesian Philosophy of Science* that a treatment of history and philosophy of science issues alongside statistical issues should be 'mutually enriching' (*ibid*.: xi).

Jaynes himself was less sure about how far one could take the historical reconstructions of scientific inference down a Bayesian route. After his discussion of Pólya's attempt to quantify Neptune discovery he claims:

But the example also shows clearly that in practice the situation faced by the scientist is so complicated that there is little hope of applying Bayes' theorem to give quantitative results about the relative status of theories. Also there is no need to do this, because the real difficulty of the scientist is not in the reasoning process itself; his common sense is quite adequate for that. The real difficulty is in learning how to formulate new alternatives which fit better the facts. Usually, when one succeeds in doing this, the evidence for the new theory soon becomes so overwhelming that nobody needs probability theory to tell him what conclusions to draw. (Jaynes forthcoming, ch. 5: 17)

This note occurs in a chapter entitled 'Queer Uses of Probability', by which he intends that at present we have no rational means for ascribing priors. So, despite his professed debt to *Mathematics and Plausible Reasoning*, we find two poles of Bayesianism represented by Jaynes and Pólya. For Jaynes, any two rational agents possessing the same information will assign identical probability functions. For Pólya, two experts with the same training may accord different changes to their degrees of belief on discovery of the same fact. One imagines a machine making plausible inferences, the other emphasises the human aspect.

Jaynes:

instead of asking, 'How can we build a mathematical model of human common sense?' let us ask, 'How could we build a machine which would carry out useful plausible reasoning, following clearly defined principles expressing an idealized common sense?' (Jaynes forthcoming, ch. 1: 5)

Pólya:

A person has a background, a machine has not. Indeed, you can build a machine to draw demonstrative conclusions for you, but I think you can never build a machine that will draw plausible inferences. (Pólya 1954b: 116)

Perhaps it is the lack of exactitude which steers Jaynes away from modelling scientific reasoning. After a lifetime investigating how symmetry considerations allow the derivation of the principles of statistical mechanics, it must be difficult to adapt to thinking about plausibility in complex situations of hypothesis assessment.

But if a physicist may be excused, what of a philosopher? John Earman, while discussing how a physicist's degrees of belief in cosmological propositions were affected by the appearance of General Relativity on the scene, tells us:

But the problem we are now facing is quite unlike those allegedly solved by classical principles of indifference or modern variants thereof, such as E. T. Jaynes's maximum entropy principle, where it assumed that we know nothing or very little about the possibilities in question. In typical cases the scientific community will possess a vast store of relevant experimental and theoretical information. Using that information to inform the redistribution of probabilities over the competing theories on the occasion of the introduction of the new theory or theories is a process that is, in the strict sense of the term, *a*rational: it cannot be accomplished by some neat formal rules, or, to use Kuhn's term, by an algorithm. On the other hand, the process is far from *ir*rational, since it is informed by reasons. But the reasons, as Kuhn has emphasized, come in the form of persuasions rather than proof. In Bayesian terms, the reasons are marshalled in the guise of plausibility arguments. The deployment of plausibility arguments is an art form for which there currently exists no taxonomy. And in view of the limitless variety of such arguments, it is unlikely that anything more than a superficial taxonomy can be developed. (Earman 1992: 197)

There seems to be no expectation here that even a qualitative Bayesianism can help us go beyond such a 'superficial taxonomy', a rather pessimistic analysis for a professed Bayesian. Does the 'limitless variety' of these arguments mean that we should not expect to find patterns among them? Despite the talk of their deployment being an 'art form', Earman does allow himself to talk about the objective quality of these plausibility arguments. Indeed, he claims that:

Part of what it means to be an "expert" in a field is to possess the ability to recognize when such persuasions are good and when they are not. (Earman 1992: 140)

Interestingly, it is Pólya the 'expert' in mathematics who believes that it is possible to extract the patterns of good plausibility arguments from his field.

So, out of the three, Jaynes, Pólya and Earman, representatives of three different types of Bayesianism, it is Pólya who believes one can say something quite concrete about plausible reasoning. All realise that plausible reasoning is a very complex process. Neither Jaynes nor Earman can see a way forward with plausible scientific reasoning. This leaves Pólya who gets involved with real cases of (his own) mathematical reasoning, which he goes on to relate to juridical reasoning and reasoning about one's neighbour's behaviour. Is he right to claim that mathematics provides a better launch pad to tackle everyday reasoning than does science?

If we want a fourth Bayesian to complete the square, we might look to the computer scientist Judea Pearl. Like Pólya, Pearl believes we can formulate the principles of everyday common sense reasoning, and like Jaynes he thinks Bayesian inference can be conducted algorithmically. To be able to do the latter requires a way of encoding prior information efficiently to allow Bayesian inference to occur. For Pearl (2001) humans typically store their background information in the form of causal knowledge. The representation of this causal knowledge in a reasonably sparse Bayesian network is the means by which a machine can be made to carry out plausible reasoning and so extend our powers of uncertain reasoning.

In earlier work (Pearl 1988) expressed his appreciation of Pólya's ideas, and yet found fault with his restriction to the elucidation of *patterns* of plausible reasoning rather than a *logic*. He considers Pólya's loose characterisation of these patterns not to have distinguished between evidence factoring through consequences and evidence factoring through causes. For instance, Pólya asserts that when B is known to be a consequence of A, the discovery that B holds makes it more likely that A holds. But this relies on the assumption, which he fails to make explicit, that B is all one has learnt. Otherwise, this pattern constitutes a well-known fallacy of causal reasoning. I see that the sprinkler on my lawn is running and that the grass is wet, but this does not make it more probable to me that it has rained recently even though wet grass is a consequence of it having done so.

So, Pólya seems to have overlooked the possibility that there may be more than one known potential cause of an event. Now, we might want to say that this is due to his working in mathematics where casual considerations are not at stake. However, one need not remain with causal stories to illustrate this fallacy. A consequence of a natural number being divisible by four is that it is even. I find that a number I seek is either 2 or 6. Although I have learnt that it is even, this discovery reduces the probability of its being divisible by 4 to zero. Essentially, what Pólya overlooked was the web-like nature of our beliefs, departing from patterns involving two propositions only when he considered the possibility of two facts having a common ground. In Bayesian networks, converging arrows are equally important but must be treated differently.

A more daring response to the claim that causes are not operative in mathematics is to deny it. In the Hellenistic era, when a broader conception of cause prevailed, this would not seem unreasonable. Proclus remarks that 'Many persons have thought that geometry does not investigate the cause, that is, does not ask the question "why?"', but then argues that they are mistaken (Morrow 1970: 158–9). As Mancosu (1996) explains, the notion of cause was also at play in seventeeth-century mathematics. Today, mathematicians are happy to say that the existence of the octonions is what *causes* the periodicity of the homotopy groups of $O(\infty)$, the inductive limit of $O(n)$, or that the reason that there are exceptional Lie groups is *because* the covering group of $SO(8)$ has an outer automorphism. More generally, when you are studying a mathematical object which displays a certain property, it is very common to wonder which features of the object are 'responsible for' that property.

It remains to be seen whether the techniques of Bayesian networks may illuminate scientific and mathematical inference. Now we shall turn our attention to examine what Bayesianism has to say about certain aspects of mathematical reasoning.

5.4 WHAT MIGHT BE ACHIEVED BY BAYESIANISM IN MATHEMATICS

Varieties of mathematical evidence may be very subtle, lending support to Earman's and Jaynes's scepticism. Pólya (1954b: 111) himself had the intuition that two mathematicians with apparently similar expertise in a field might have different degrees of belief in the truth of a result and treat evidence for that result differently. Even though each found a consequence of the result equally plausible, the establishment of this consequence could have an unequal effect on their ratings of the likelihood of the first result being correct. The complex blending of the various kinds of evidence experienced through a mathematician's career would explain the differences in these reactions, some of which might be attributable to aspirations on the part of each of them either to prove or disprove the theorem. But Pólya goes further to suggest that such differences of judgement are based on 'still more obscure, scarcely formulated, inarticulate grounds' (*ibid.*).

Certainly, evidence for the correctness of a statement may be very subtle. It may even arise through an experience of failure. In chapter 7 we shall see that, while proving the so-called 'duality theorem', Poincaré had come to realise that an assumption he was making about the way differential manifolds intersect was invalid in general. However, he still believed that the general strategy of constructing for a given set of manifolds of equal dimension a manifold of complementary dimension which intersected each of the members of the set exactly once could be made to work. He just needed to have the intersections occur in a more controlled fashion. One can only guess how this experience impacted on his degree of belief in the duality theorem. It is quite probable that even though the initial proof was found to be wrong, the experience of near success with a variant of a strategy gave him hope that another variant would work. It must also happen, however, that mathematicians are discouraged by such setbacks.

Evidence can also involve the non-discovery of something, as Sherlock Holmes well knew when he built his case on the observation of a dog that did not bark. The classic example of the unsurveyable human-generated kind of proof at the present time is the classification of finite simple groups into 5 infinite families and 26 sporadic outsiders. How does one's degree of belief in this result depend on such potentially flawed lengthy evidence? Fallis (1997) has Gorenstein, the driving force behind the collective proof, confessing that confidence is boosted less by the proof itself than by the fact that no other such groups have been found. Similarly, remarks are often to be heard concerning the consistency of ZFC that one would have expected to have encountered a contradiction by now.

We should also remember that evidence for mathematical propositions comes from sources which have only recently become available, as for example in the case of computers. Reliance on computer evidence raises some novel issues. Oscar Lanford is attributed with pointing out that:

> in order to justify a computer calculation as part of a proof..., you must not only prove that the program is correct (and how often is that done?) but you must understand how the computer rounds numbers, and how the operating system functions, including how the time-sharing system works. (Hirsch 1994: 188)

Moreover, if more than one piece of computer evidence is being considered, how do we judge how similar they are for conditionalising purposes? This would require one to know the mathematics behind any similarities between the algorithms utilised.

It is clear then that any account of mathematical inference will require a very expressive language to represent all the various forms of evidence which impact on belief in mathematical propositions. The Bayesian wishing to treat only propositions couched in the language of the object level might hope to be able to resort to Jeffrey conditionalisation, but this comes at the price of glossing over interesting features of learning. Concerning scientific inference, Earman (1992: 196–8) asserts that many experiences will cause the scientist to undergo what he calls *non-Bayesian* shifts in their degrees of belief, i.e., ones unaccountable for by any form of algorithmic conditionalisation. These shifts, the resetting of initial probabilities, are very common, he claims, arising from the expansion of the theoretic framework or from the experience of events such as '[n]ew observations, even of familiar scenes; conversations with friends; idle speculations; dreams' (1992: 198).

One might despair of making any headway, but taking Pólya as a guide we may be able to achieve something. While recognising that making sense of plausible reasoning in mathematics will not be easy, I believe that three key areas of promise for this kind of Bayesianism in mathematics are analogy, strategy and enumerative induction.

Analogy

As we saw in chapter 3, before turning to a probabilistic analysis of plausible reasoning in the second volume of *Mathematics and Plausible Reasoning*,

Pólya had devoted the first volume (1954a), as its subtitle suggests, to the themes of analogy and induction. Now, analogies vary as to their precision. When vague they contribute to what he called the *general atmosphere* surrounding a mathematical conjecture, which he contrasts to pertinent *clear facts*. While verifications of particular consequences are straightforwardly relevant facts, the pertinence of analogical constructions may be hard to discern precisely.

Let us illustrate this with an example. At the present time, the vast majority of mathematicians have a high degree of belief in the Riemann Hypothesis. Recall that the Riemann zeta function is defined as the analytic continuation of $\zeta(s) = \Sigma n^{-s}$ summed over the natural numbers, and that the hypothesis claims that if s is a zero of $\zeta(s)$, then either $s = -2, -4, -6, \ldots$, or the real part of s equals $1/2$. Many roots have been calculated (including the first 1.5 billion zeros in the upper complex plane along with other blocks of zeros), all confirming the theory, but despite this 'overwhelming numerical evidence, no mathematical proof is in sight' (Cartier 1992: 15). As Bayesians have explained, there are limits to the value of showing that your theory passes tests which are conceived to be very similar. If, for example, a further 100 million zeros of the zeta function are found to have their real part equal to $1/2$, then little change will occur in mathematicians' degrees of belief, although a little more credibility would be gained if this were true of 100 million zeros around the 10^{20}th, which is precisely what has happened.

In this example the clear facts making up the numerical evidence can lend only limited credence by themselves. After all, there are 'natural' properties of the natural numbers which are known to hold for exceedingly long initial sequences. What counts in addition beyond evidential facts, however numerous, is the credibility of stronger results, general consequences and analogies. Indeed, if an analogy is deemed strong enough, results holding for one side of it are thought to provide considerable support for their parallels. Concerning the Riemann conjecture (RC), we are told that:

There is impressive numerical evidence in its favour but certainly the best reason to believe that it is true comes from the analogy of number fields with function fields of curves over finite fields where the analogue of RC has first been proved by A. Weil. (Deninger 1994: 493)

This analogy,[6] we saw in chapter 4, was postulated early in this century as a useful way of providing a halfway house across Dedekind and Weber's older

---

[6] See also Katz and Sarnak (1999), in particular the table on p. 12.

analogy from algebraic number fields to function fields over the complex numbers. The more geometric side of the analogy Deninger mentions was able to absorb cohomological techniques, allowing Weil to prove the Riemann hypothesis analogue in 1940. An extraordinary amount of effort has since been expended trying to apply cohomology to number theory (Weil, Grothendieck, Deligne, etc.) with the establishment of the standard Riemann hypothesis as one of its aims.

How should we judge how analogous two propositions, A and B, are to each other? For Pólya (1954b: 27) it correlates to the strength of your 'hope' for a common ground from which they both would naturally follow. Of course, how this hope is to be assessed will depend on your sense of the relatedness of mathematical facts. This sense seems to vary between individual mathematicians, and also between communities of mathematicians. The effect of the success of the Dedekind–Weber analogy could only have been to strengthen the idea of mathematics' connectivity and so to have encouraged the hope for common ground in a wider range of situations.

With a conjectured common ground H in place, Pólya argues, an increase in confidence in A will then feed up to it and then back down to B.[7] Recall from chapter 4 that in Pólya's principal example, Euler noticed that the function $\sin x/x$ resembles a polynomial in several respects: it has no poles; it has the right number of zeros, which do not accumulate; it behaves symmetrically at $\pm\infty$. On the other hand, unlike a polynomial, $\sin x/x$ remains bounded. Even with this disanalogy, it seemed plausible that $\sin x/x$ shared enough of the properties of polynomials that it would possess those 'responsible for' allowing factorisation in the case of the latter, giving him the confidence to try 'factorising' the Taylor expansion.

It might be that what is happening here is something similar to what Pearl (2000) has termed the 'transfer of robust mechanisms to new situations'. We have a mechanism that links factorisation of a function to its zeros. We find it applies for complex polynomials and wonder whether it may be extended. Features of polynomials that may be required in the new setting are that they have the right number of zeros, they remain bounded on compact sets, and they behave similarly at $\pm\infty$. Might the mechanism be expected to work for a non-polynomial function possessing these features, such as $\sin x/x$? What if you force the variable measuring the number of roots to be infinite? We may find it hard to estimate quantitatively the similarity between a function like $\sin x/x$ and a complex polynomial, but

---

[7] Notice here the flavour of a Bayesian network: H pointing to both A and B.

it is clear that tan x/x or exp x are less similar, the former having poles, the latter having no zeros and asymmetric behaviour at $\pm\infty$, and indeed the mechanism does fail for them.

In this case, once the realisation that an analogy was possible, it did not cost much to work through the particular example. Euler hardly needed to weigh up the degree of similarity since calculations of the sum quickly convinced him that sin x/x could be factorised. However, to develop a general theory of the expansion of complex functions did require greater faith in the analogy. It paid off when further exploration into this mechanism permitted mathematicians to form a very general result concerning entire complex functions, providing the 'common ground' for the analogues.

Strategy

Moving on now to strategy, the title of Deninger's (1994) paper – 'Evidence for a Cohomological Approach to Analytic Number Theory' – is also relevant to us. His aim in this paper is to increase our degree of belief that a particular means of thinking about a field will lead to new results in that field. This is a question of strategy. At a finer level one talks of tactics. Researchers from the AI community working on automated theorem proving, have borrowed these terms. In chapter 2, we mentioned the tactic devised by Larry Wos (Wos and Pieper 1999) which involves thinking in terms of how probable it is that the computer can reach the target theorem from a particular formula generated from the hypotheses during the running of the program. This tactic takes the form of a weighting in the search algorithm in favour of formulas which have a syntactical form matching the target.

Elsewhere, researchers in Edinburgh have been interested in the idea of the choice of tactics (Bundy 1999). There is an idea of likening mathematics to a game of bridge where the mathematician, like the declarer, has some information and a range of strategies to achieve their goal (finesse, draw trumps, squeeze). Of course, there is a difference. In bridge, you are in the dynamic situation where you cannot try out every strategy, as the cards get played. This forces you to pay very close attention to which tactics have the best chance of working. In mathematics, on the other hand, with a computer it does not cost you much to try things out, although one does risk combinatorial explosion.

At present, although probabilities are being used by the Edinburgh group for their computer bridge player, they are not yet being used for their automated theorem provers. While the computer has a small repertoire of syntactical tactics (rippling, resonance, heat, etc.) there is less need for an assessment of the chance of each working, but presumably the number of proof techniques will grow.

I mentioned at the end of chapter 4 that Gowers (2000b) describes a fantastic proof assistant, a kind of surrogate supervisor, able to suggest proof techniques based on extensive experience of what tends to work in specific problem situations, and vast knowledge of the literature. If such a device ever comes to be built, we might foresee a role for Bayesianism, and its constructors will no doubt have Pólya to thank. To give a brief flavour of his ideas, when planning to solve a problem, any of the following should increase your confidence in your plan (Pólya 1954b: 152–3):

(1) Your plan takes all relevant information into account.
(2) Your plan provides for a connection between what is known and what is unknown.
(3) Your plan resembles some which have been successful in problems of this kind.
(4) Your plan is similar to one that succeeded in solving an analogous problem.
(5) Your plan succeeded in solving a particular case of the problem.
(6) Your plan succeeded in solving a part of the problem.

Whether automated reasoning devices can prosper by simulating these considerations in a probabilistic framework remains to be seen.

Enumerative induction

Besides the incorrect Bayesian calculation of the confirmation provided by the observation of Neptune, Pólya does resort to a quantitative sketch in another place (1954b: 96–7). Here he outlines how one might think through the boost to the credibility of Euler's formula for a polyhedron (vertices – edges + faces = 2) known to hold for some simple cases, when it is found to be true of the icosahedron. $(12 - 30 + 20 = 2)$. Pólya's approach is to reduce the problem to the chances of finding three numbers in the range 1 to 30 with the property that the second is equal to the sum of the other two, i.e., $(V - 1) + (F - 1) = E$. The proportion of these triples is around 1 in 60, providing, Pólya argues, a boost of approximately 60 to the prior probability of Euler's conjecture. Here again we see the same problem that Jaynes located in the Neptune calculation. The ratio of the likelihood of the Euler conjecture compared to that of its negation is 60.

In any case Pólya's construction can only be viewed as sketchy. It is not hard to see that the number of edges will always be at least as great as one and a half times the number of faces or the number of vertices. (For the latter, for example, note that each edge has two ends, but at least three of these ends coincide at a vertex.) Thus one should have realised that there are further constraints on the possible triples and hence that the likelihood ratio due to the evidence for the Euler formula should have been in comparison to better informed rival conjecture, and so not so large. But the interesting point is that Pólya goes on to say that:

If the verifications continue without interruption, there comes a moment, sooner or later when we feel obliged to reject the explanation by chance. (1954b: 97)

The question then arises as to whether one is justified in saying such a thing on the basis of a finite number of verifications of a law covering an infinite number of cases. This will hinge on the issue of the prior probability of such a law.

Now, consider Laplace's rule of succession. If you imagine yourself drawing with replacement from a bag of some unknown mixture of white and black balls, and you have seen m white balls, but no black balls, the standard use of the principle of indifference suggests that the probability that the next n will be white is

$$(m + 1)/(m + n + 1).$$

As $n \to \infty$, this probability tends to zero. In other words, if verifying a mathematical conjecture could be modelled in this fashion, no amount of verification could help you raise your degree of belief above zero.

This accords with the way Rosenkrantz (1977) views the situation. He considers the particular case of the twin prime conjecture: that there are an infinite number of pairs of primes with difference 2. He mentions that beyond the verification of many cases, there are arguments in analytic number theory which suggest that you can form an estimate for the number of twin primes less than n and show that it diverges. He then continues:

Now if Popper's point is that no examination of 'positive cases' could ever raise the probability of such a conjecture to a finite positive value, I cannot but agree. Instances alone cannot sway us! But if his claim is that *evidence of any kind* (short of proof) can raise the probability of a general law to a finite positive value, I emphatically disagree. On the cited evidence for the twin prime conjecture, for example, it would seem to me quite rational to accept a bet on the truth of the conjecture at odds of, say 100:1, that is to stake say $100 against a return of $10 000 should the conjecture prove true. (Rosenkrantz 1977: 132)

So for Rosenkrantz, with no background knowledge, the principle of indifference forces a universal to have zero, or perhaps an infinitesimal (something also considered by Pólya) prior probability. However, other considerations may determine a positive probability:

Subject-specific arguments usually underlie probability assessments in mathematics. (Rosenkrantz 1977: 90)

In support of this view, returning to the Euler conjecture, we should note that there was background knowledge. For polygons, it is a trivial fact that there is a linear relation between the number of vertices and the number of edges, namely, V = E. Hence, a simple linear relation might be expected one dimension higher.

But is it always this kind of background knowledge which gives the prior probability of a conjecture a 'leg-up'? Do we ever have a situation with no background knowledge, i.e., where a *general atmosphere* is lacking? Consider the case of John Conway's 'Monstrous Moonshine', the conjectured link between the *j*-function and the monster simple group. The *j*-function arose in the nineteenth century from the study of the parameterisation of elliptic curves. It has a Fourier expansion in $q = \exp(2\pi i \tau)$:

$$j(\tau) = 1/q + 744 + 196884q + 21493760q^2 + 864299970q^3 + \cdots$$

One day while leafing through a book containing this expansion, a mathematician named John MacKay observed that there was something familiar about the third coefficient of this series. He recalled that 196,883 was the dimension of the smallest non-trivial irreducible representation of what was to become known as the monster group, later confirmed to be the largest of the 26 sporadic finite simple groups. Better still, adding on the one dimension of the trivial representation of the monster group results in equality.

In view of the very different origins of these entities, the *j*-function from nineteenth-century work on elliptic curves and the monster group from contemporary work in finite group theory, if one had asked a mathematician how likely she thought it that there be some substantial conceptual connection between them or common ground explaining them both, the answer would presumably have been 'vanishingly small'. In Bayesian terms, Pr(connection | numerical observation) is considerably greater than Pr(connection), but the latter is so low that even this unlikely coincidence does not bolster it sufficiently to make it credible. Naturally, McKay was told that he was 'talking nonsense'. He then went on, however, to observe that the second non-trivial representation has dimension 21296876. A quick calculation revealed that the fourth coefficient of the *j*-function could be expressed as: 21493760 = 21296876 + 196883 + 1. In fact every further coefficient of the j-function turns out to be a simple sum of the dimensions of the monster's representations. At this point the question of whether there is some connection has been all but answered – it has become a near certainty. Conway challenged the mathematics community to resolve this puzzle:

Fourier expansion in $q = \exp(2\pi i \tau)$:

$$j(\tau) = 1/q + 744 + 196884q + 21493760q^2 + 864299970q^3 + \cdots$$

196884	$196883 + 1$
21493760	$21296876 + 196883 + 1$
864299970	$842609326 + 21296876 + 196883 + 196883 + 1 + 1$
$\cdots$	$\cdots$

The answer eventually arrived through a construction by Richard Bocherds, a student of Conway, which earned him a Fields' Medal. Bocherds managed to spin a thread from the j-function to the 24-dimensional Leech lattice, and from there to a 26-dimensional space-time inhabited by a string theory whose vertex algebra has the monster as its symmetry group.

So why does the monster group–j-function connection become so likely by the time you have seen three or four of the sums, even with a minuscule prior, when other inductions are less certain after billions of verifications? Would we find consensus on how the number of instances affects one's confidence? Surely most people would agree that it was a little reckless on Fermat's part to conjecture publicly that $2^{2^n} + 1$ is prime after verifying only five cases (and perhaps performing a check on divisibility by low primes for the sixth).

n	0	1	2	3	4	5
$2^{2^n} + 1$	3	5	17	257	65537	$4294967297 = 641 \times 6700417$

But perhaps mathematicians were less cautious in a time when discovered patterns more often than not were found to be valid everywhere. Pólya (1941: 456) quotes Descartes: 'In order to show by enumeration that the area of a circle is greater than that of any figure of the same perimeter, we do not need to make a general investigation of all the possible figures, but it suffices to prove it for a few particular figures whence we can conclude the same thing, by induction, for all the other figures.'

Is it possible to use Bayes' theorem, even merely suggestively? Let us return to the case of the Riemann hypothesis (RH). If we have a prior degree of belief for RH, how can 1.5 billion verifications affect it? Surely they must, but then is there some asymptotic limit? One might choose to factor the posterior for RH as follows

```
Pr(RH|Data) = Pr(RH|p = 1, Data)·Pr(p = 1|Data),
```

where p denotes the limiting proportion, if this exists, of the zeros that lie on the line, taking the zeros in the order of increasing size of modulus.

For the second factor we might then have started with a prior distribution over p according to the weighted sum of the exhaustive set of hypotheses about p: non-convergent p; p in $[0,1)$; $p = 1$.[8] Then if one can imagine some element of independence between the zeros, e.g., the fact that the nth zero lies on the line provides no information on the $(n + 1)$th, then the confirmation provided by the 1.5 billion zeros should push the posterior of $p = 1$ to take up nearly all the probability accorded to convergent p. This kind of assumption of independence has been used by mathematicians to make conjectures about the distribution of primes, so may be appropriate here. In Bayesian terms, it would be better to use de Finetti's idea of exchangeability, that is, we have no reason to prefer any particular ordering of a given number of successes and failures. We might also consider that 1.5 billion positive instances provides an indication that p is convergent. Again, however, this consideration would depend on experience in similar situations.

For the first factor, out of all the functions you have met for which their zeros have held for a large initial section and the proportion of cases is 1, you are wondering what proportion are universally true. It is clear, then, that again much would depend on prior experience. For example, something that would be kept in mind is that the function $\pi(x)$, defined as the number of primes less than x, is known to be less than a certain function, denoted $\mathrm{li}(x)$, up to 10^{12}, and that there is good evidence that this is so up to 10^{30}. But it is known not to hold somewhere before 10^{400}. Indeed, there appears to be a change close to 1.4×10^{316}.

Returning finally to 'Monstrous Moonshine', perhaps we should look harder for a reliance on background knowledge to explain the rapid appearance of conviction. First, it is worth remembering that the dimensions of the monster group's representations and the coefficients of the j-function were not 'made up'. They come from 'natural' mathematical considerations, a term we shall discuss in chapter 9. Imagine in the Monstrous Moonshine

---

[8] Levinson showed in 1974 that $p > 1/3$.

case if the two sides were not 'interesting' entities or that you knew for a fact that these numbers were randomly generated, wouldn't you take more convincing? Similar considerations are discussed by Paris *et al.* (2000), who wish to justify some 'natural' prior distribution of probability functions over *n* variables:

> what in practice I might claim to know, or at least feel justified in believing, is that the data I shall receive will come from some real world 'experiment', some natural probability function; it will not simply have been made up. And in this case, according to my modeling, I do have a prior distribution for such functions. (Paris *et al.* 2000 : 313)[9]

Evidence for the fact that background knowledge is coming into play in this case is provided by the fact that on presentation of the example to an audience of non-mathematicians they found the numerical coincidences not at all convincing. Despite the fact that a mathematician has no knowledge of a reason for a connection between these two mathematical entities, some slight considerations must play a role. Indeed, what seemed to disappoint the non-mathematicians was the need to include multiples of the dimensions of the irreducible representations. A mathematician, on the other hand, is well aware that in general a group representation is a sum of copies of irreducible ones. For example, the right regular representation, where the group acts on a vector space with basis corresponding to its own elements, is such a sum where the number of copies of each irreducible representation is equal to its dimension. Behind the addition of dimensions are sums of vector spaces. Second, a mathematician would know that the *j*-function arises as a basic function, invariant under the action of the modular group. This offers the possibility that group theory might shed some light on the connection.

5.5 CONCLUSION

We have covered a considerable stretch of ground here. Clearly much work remains to be done on Pólya's research programme, but I think we can allow ourselves a little more optimism than Earman shows towards philosophy of science. I have isolated the following areas as potentially useful to study in a Bayesian light: (1) Analogy; (2) Strategy choice; and, (3) The use of large computations to increase plausibility of conjectures. In chapter 6 I shall

---

[9] The degree to which an object is deemed natural correlates to expectations about how it may be integrated into the rest of mathematics: 'Now, the Monster M is presumably a natural mathematical object, so we can expect that an elegant construction for it would exist' (Gannon 2001: 8).

be considering an additional area (4) Mathematical predictions in physics; and, elsewhere, (5) The use of stochastic ideas in mathematics (random graphs, random matrices, etc.). It is important to note that we need not necessarily arrive at some quantitative, algorithmic Bayesian procedure to have made progress. If Bayesianism in mathematics suggests interesting questions in the philosophy of mathematics, then I think we can say that it has served its purpose.

It is all too typical that Pólya's important contribution to Bayesianism has been overlooked by philosophers. Inspired by de Finetti he, in turn, inspired Jaynes and Pearl with the ideas he had worked out by thinking about mathematical reasoning. Meanwhile, philosophers happy to accept Hempel's (1945) claim that mathematics is just a branch of logic, have passed over Pólya's work. Let's now see how this mathematical blind spot operates in the philosophy of science.

CHAPTER 6

Uncertainty in mathematics and science

nothing is evidence except in relation to some definite question.

(Collingwood 1999: 37)

6.1 INTRODUCTION

We have seen in chapter 5 that to model plausible mathematical reasoning in Bayesian terms it is best to forgo logical omniscience. Now, the Bayesian philosopher of science might be quite happy to do so in purely mathematical contexts, while at the same time maintaining that omniscience is not an overly inaccurate assumption for what concerns her – the treatment of confirmation in the natural sciences. Indeed, she might argue that with so much uncertainty surrounding the empirical adequacy of our models, it would be fastidious to worry about our mathematical imperfections. If you put ten male rabbits into a pen with ten female rabbits and come back a few weeks later expecting there to be twenty rabbits but count twenty-five, then you do not question your arithmetic beliefs, but rather your counting and then your model of the stability of the number of rabbits over time. But might this idealisation make us miss something about scientific reasoning itself, especially when our mathematics is not secure? After all, some pieces of physics we trust more than we do some pieces of mathematics. Think of having made a conjecture in combinatorics which you have verified on paper in a few simple cases. You set your computer algebra system onto the problem, and it duly carries out several more calculations which provide confirmation for your conjecture. At this point you may be quite confident of its truth. Eventually, however, you encounter a counter-example, and repeat the calculation on a different machine using what you know to be a different computer algebra package. Naturally, you give up your conjecture sooner than believe that both hardware–software systems have led to the same mistake, and you certainly will not take it as a refutation of the semiconductor theory used in the design of the hardware.

Some more 'humanistic' versions of Bayesianism do already allow for scientists conditionalising on the basis of the discovery of 'logico-mathematical consequences' of a theory.[1] Then not only do we talk of $Pr(T)$ and $Pr(T|E)$, but also of $Pr(T \vdash E)$ and $Pr(T | T \vdash E)$. This extension is generally considered in situations where some already observed phenomenon remains unexplained until a theory, proposed for whatever reason, is found at some later date to account for it, using a piece of established mathematics. My concern, however, is that the notion of a theory accounting for evidence underpinning entities of the form $Pr(T \vdash E)$ has not been satisfactorily described, especially in cases where the relevant mathematics has yet to be devised. What is overlooked here is the fact that in many cases it is necessary that a large amount of mathematical machinery be produced to allow the production of a model through which a scientific theory can be said to explain or predict a particular phenomenon, and that it may happen that over long periods of time parts of the mathematics are uncertain. A physicist may come to realise that were a certain mathematical result true, she could establish that an observation is accounted for by her pet theory. Thus she makes a prediction that the mathematics is true and may even be able to give a heuristic argument for its truth from her understanding of the physics. If the mathematical result turns out to be correct, credit accrues to the scientific theory, the more so the more unlikely the result. To portray this kind of situation as merely the elaboration of the mathematical content of that scientific theory, as though the mathematical component of scientific activity consists in little more than the uncovering of what is already contained within scientific precepts, is a serious misrepresentation.

I shall argue in this chapter that *any* theory of scientific inference needs to take the mathematical reasoning fully into account, but to make the discussion more concrete I shall focus on plausibility considerations and continue speaking in Bayesian terms. I take it as fairly obvious that a Bayesian philosopher of science wishing to take uncertain mathematical knowledge into consideration will construe rationality in mathematics in Bayesian terms. In section 6.2 I shall present the cases of four less visited episodes from the history of science, illustrating how uncertainty in mathematics and physics may affect each other. In section 6.3, I analyse the ways in which one should construe confirmation as taking place, drawing on recent philosophical work on scientific models. Then, in section 6.4, on the basis of this analysis, I shall outline how a Bayesian philosopher of science might resolve the infamous 'old evidence' problem.

---

[1] A discussion of the relevant literature may be found in ch. 5 of Earman 1992.

6.2 FOUR CASE STUDIES OF THE INTERRELATIONSHIP BETWEEN THE PLAUSIBILITY OF SCIENTIFIC HYPOTHESES AND EVIDENCE AND THE PLAUSIBILITY OF MATHEMATICAL CONJECTURES

The stability of the solar system

The question as to whether our solar system will persist in a stable planetary configuration has stimulated thinkers for over 2,000 years. Of course, Greek understanding of the problem was very different from ours, stability of the heavens being assured by the location of the celestial bodies on crystal spheres. A formulation of the problem closely comparable to a present-day one came only with the publication of Newton's *Principia*. But despite the power of his mathematical approach to cosmology, it was soon realised that many aspects of planetary dynamics would be difficult to treat. Newton himself worked on the three-body system, comprising the sun, the earth and the moon, and so began the study of the *three-body problem*.

Insight into many-body dynamics was difficult to acquire, although notable successes did occur. In 1785 Laplace took a major step in promoting Newtonian gravitation when he demonstrated that it could explain perturbations in the orbits of Jupiter and Saturn as arising from the near-resonance in the frequency of their orbits. He showed how the 2:5 ratio of their orbital periods caused deviations in their elliptical orbits having a 900-year period. From experience with this and other aspects of celestial mechanics, Laplace was convinced of planetary stability and he wrote in the preface to volume 3 of *La Mécanique Céleste*:

Nature orders the celestial machine for an eternal duration, upon the same principles which prevail so admirably upon the earth, for the preservation of individuals and for the perpetuity of the species.

Through the nineteenth century numerous researchers attempted analytic solutions of various versions of the many-body problem. A question on this subject was posed in a prize competition offered by King Oscar of Sweden, which, as June Barrow-Green (1997) has shown, was rigged in favour of the French mathematician Henri Poincaré. By the time he had been allowed to correct his entry, Poincaré had established a new qualitative approach to the treatment of dynamical systems, and over the following years he developed many essential components of the branch of mathematics it needed. One of the outcomes of Poincaré's work was a demonstration that a certain series expansion, thought previously to converge, in fact diverges. With this and the first recognition that dynamical systems may possess what we now term *chaos*, confidence in the stability of the solar system diminished.

Significant progress on the stability of the solar system had to wait until *KAM theory* was developed in the middle of the twentieth century. In dynamical systems theory, the evolution of a system is seen as a point tracing out a path in a phase space. In conservative systems energy conservation restricts this path to lie on some bounded subspace, a many-dimensional torus, parameterised by a set of numbers. KAM theory studies the stability of these tori when a small perturbing force is added, where interplanetary attractions play this role in the case of the solar system. It demonstrates that tori whose parameters are sufficiently far from being dependent over the integers will survive the perturbation, merely being deformed. If KAM theory is relevant to a system, it implies what Arnold (1988: 189) describes as metric stability, but topological instability or, in other words, stability for the majority of initial conditions, yet in however small a neighbourhood of the initial conditions there passes an unstable trajectory. If applicable to the solar system it suggests that it is overwhelmingly likely that the planets' positions will remain in a bounded region of space, but that some almost identical solar system to ours would be unstable. To complicate matters mathematicians have devised several other notions of stability as applied to dynamical systems.

A proof of the central theorem of KAM theory in the case of two degrees of freedom was sketched in lectures by Kolmogorov in 1954. In 1963 Arnold constructed a proof for systems with any number of degrees of freedom. Moser later showed that some of the conditions on the differentiability of the Hamiltonian governing their dynamics could be relaxed. Arnold's assessment at the time of its relevance to the stability question was that:

> For the majority of initial conditions under which the instantaneous orbits of the planets are close to circles lying in a single plane, perturbation of the planets on one another produces, in the course of an infinite interval of time, little change on these orbits provided the masses of the planets are sufficiently small. (Arnold 1963: 125)

Clearly there had been a boost to belief in stability in relation to the time earlier in the century when mathematicians, such as Birkhoff, had 'inclined to the opposite view' (*ibid.*). This boost might typically have occurred gradually from the moment when Kolmogorov conjectured KAM theory and had good reason to believe it, to the moment when Moser had demonstrated the theorem for a broad range of systems. A mathematician's degree of belief in KAM theory would have fluctuated through the decade as promising lines of attack on the problem came and went, and so the Bayesian might calculate the degree of belief in the stability of the solar system as varying via

$$\mathrm{Pr(S/NG\ \&\ K) = P(S/KAM\ \&\ NG\ \&\ K) \cdot Pr(KAM)}$$
$$+ \mathrm{Pr(S/\neg KAM\ \&\ NG\ \&\ K) \cdot Pr(\neg KAM),}$$

where S = stable solar system, NG = Newtonian gravitation (as an adequate approximation), K = background knowledge.

To describe this episode as just the drawing of a logico-mathematical consequence of a scientific theory 300 years after it had been proposed is surely misleading. Rather, the development of KAM theory constitutes the production of a result which is true of a wide range of mathematical dynamical systems and which may or may not be applicable to the dynamics of the solar system. Whether it is applicable is still to be decided:

> The rigorous bounds that have been established, so far, which guarantee the existence of some invariant tori, do not come close to establishing the stability of the major planets in the Solar System. There is a gap between the bounds that are proved and experimental evidence on how far they are valid. (Lagarias 1992: 39)

The factors involved in determining what would be denoted $\mathrm{Pr(NG \vdash S)}$ by the so-called *humanistic* Bayesian are subtle indeed.

Although KAM theory provided support for stability, the clearest idea we have today of the future of the solar system is based on computer simulations of its evolution. It seems that the future is uncertain. The greatest causes for alarm are the possible drifting away of Mercury, the transport of asteroids to Mars, and even to the Earth, owing to a 3:1 resonance between Jupiter's orbit and their own, and the chaotic change in the eccentricity of Mars which may allow it within 5 billion years to cross the path of the Earth's orbit.[2]

Computer evidence raises a host of epistemological questions. Indeed, the use of computer simulations and sampling techniques in the biological and physical sciences provides further ports of entry for changing levels of credibility in pieces of mathematics to affect the credibility of scientific results. Concerning such evidence you will need to know the extent to which the algorithm used can be thought to have captured the system of equations in the model to set your degree of belief in the validity of using it. But to make a plausible case for this may well require novel types of mathematical demonstration. Then when it comes to different computer simulations used

---

[2] See Sussman and Wisdom (1992).

as evidence, one needs to know how 'different' are the resulting pieces of evidence, for the Bayesian, Pr(second evidence/first evidence), based on mathematical analysis of the simulation set-ups.[3] Developments in the mathematics of algorithms will effect such judgements.

However we decide to interpret the notion, at the present time, the presence of stability in the solar system in the remaining 5 billion years before the sun is expected to turn into a red giant is very uncertain. Degrees of belief in such stability have varied over the past 300 years, and at all times have depended on the credibility of uncertain pieces of mathematics in very subtle ways.

The onset of turbulence

Let us now consider a case where scientific observation provides evidence for a mathematical conjecture. In his account of chaos theory, *Does God Play Dice?*, Ian Stewart (1997) discusses Ruelle and Takens' proposed scenario for the way turbulence sets in during changes in the parameters of fluid flow. Fluid flows are modelled by Navier–Stokes equations, notoriously difficult beasts to tame mathematically. Still, models employing these equations have done tremendously well in spite of the fact that the physical assumptions behind their use, such as the continuity of the medium, clearly contradict the best account we have of physics at the smallest scales. Now, Ruelle and Takens conjectured that under certain conditions, the dynamics of a fluid determined by the Navier-Stokes equations is governed by a strange attractor. As Stewart explains:

So one possible research programme to put the Ruelle–Takens theory on a testable basis is: derive a strange attractor from the Navier–Stokes equations for fluid flow. This is a problem that requires mathematical, rather than experimental advances, and it hasn't been carried out yet. (Stewart 1997: 172)

The Navier–Stokes equations and the ways of solving, approximately solving, or qualitatively treating them belong to mathematics. Stewart's suggestion would thus appear to involve merely internal adjustments to degrees of belief in mathematical propositions. But Stewart then proceeds to explain how Mitchell Feigenbaum extracted a universal feature of the development of chaos in very simple dynamical systems. These systems included the well-known logistic mapping, a way of using a parabola to generate a dynamics, and some close relatives. The common feature concerned a certain

[3] Sussman and Wisdom tell us how convincing they find 'the detailed agreement between our 100-million-year solar system integration and that of Laskar [another modeller – DC], because of the radically different methods used' (1992 : 61).

constant 4.669201609 . . . , arising from the ratio of changes in the value in a parameter at successive points of bifurcation of the dynamics. These findings lead Stewart to the conclusion:

> if it so happens that, buried away in the Navier–Stokes equations, there's a mathematical process involving a simple-humped mapping, then a period-doubling cascade with scaling ratio 4.669 is going to occur. (Stewart 1997: 196)

What we have here then is a mathematical conjecture that there is a technique of the kind mathematicians employ to expedite the solution of differential equations (change of variables, etc.) which will lead to a process involving such a mapping. As yet there is neither proof nor any convincing mathematical evidence for this conjecture. But what if experiments on fluid flow produced Feigenbaum's constant? We would then have 'experimental evidence in favour of a mathematical theorem!' (*ibid.*). Stewart describes this situation as 'bizarre', but he goes on to tell us that this is precisely what occurred when Albert Libchaber later made observations of convection rolls in a miniscule chamber filled with liquid helium, from which he calculated a number close to Feigenbaum's constant.

The effect of this experiment is to make degrees of belief change. Prior to the experiment we would expect:

Pr(Navier-Stokes equations involve a single-humped mapping/appearance of Feigenbaum's constant in fluid flow data) > Pr(Navier-Stokes equations involve a single-humped mapping).

Thus, to rely on all possible evidence to determine her degrees of belief a mathematician will need to have degrees of belief in relevant scientific statements. The acceptance of Libchaber's experimental results should then have an impact on degrees of belief concerning the properties of other systems thought to be modelled by Navier–Stokes equations.

Notice how one might have reacted in an extended Duhem–Quine fashion had Libchaber's experiment failed. Had he arrived at a figure sufficiently different from 4.669, the response might have been:

(a) The experiment was incorrectly carried out because some auxiliary hypothesis did not hold – the correct value should have been close to 4.669

(b) Navier–Stokes modelling was wrong/inaccurate – we need some other, or at least a more refined, model of liquid helium flow

(c) The mathematical conjecture was wrong – there is no single-humped mapping involved in the Navier–Stokes equations.

It is clear I hope that if Bayesianism is to explicate all varieties of the Duhem–Quine problem, the resolution of which has been deemed an important success (Howson and Urbach 1989: 92–102), it must take into account degrees of belief in mathematical statements.

The Lee–Yang circle theorem

The use of equilibrium statistical mechanics (canonical ensembles, partition functions, etc.) to study physical macrosystems by way of the interactions of their microscopic constituents was well established by the 1950s. A long-standing, central idea of this field is that an examination of the parameter values at which the partition function fails to be analytic in the thermo-dynamic limit (e.g. as a lattice becomes infinitely large) gives information regarding phase transitions in the macrosystem. Now, ferromagnets in the presence of an external magnetic field were thought to be suitably mod-elled by a two-dimensional lattice, at the vertices of which were situated atoms of spin up or spin down, an Ising model. Below a certain temperature (the Curie temperature), it was known that ferromagnets are spontaneously magnetised and that if the external magnetic field perpendicular to the lat-tice, H, is manipulated, a sudden change in their magnetisation occurs as it passes through zero. This led to the prediction that the partition func-tion for infinitely many particles would fail to be analytic at $H = 0$ for sufficiently low temperatures. Translating this piece of physics into the mathematics of the Ising model generated the prediction that members of a certain class of complex polynomials in one variable always have their zeros lying on the unit circle. This so-called Lee–Yang circle theorem was eventually proved after some considerable difficulty.

For the mathematicians, the magnetisation data and the previous success of the equilibrium statistical mechanics machinery should have counted to set their degree of belief in the circle theorem fairly high. For the scientists, as the theorem becomes more plausible, first through some trial-and-error experimentation with these polynomials and then through the establish-ment of a proof, increases of credibility were fed back into the physics which had led to its conjecture, into the validity of the model, the experimental data and the whole statistical mechanics approach.

Quantum field theory

By the end of the twentieth century, novel relationships between mathemat-ics and physics had emerged. At times, given the nature of some modern physical theories, their mathematical predictions are the only ones with any chance of being confirmed. The mathematician Sir Michael Atiyah explains the use of such predictions in modern theoretical physics:

> quantum field theory has had its credibility enhanced by its success in making correct mathematical predictions. Given the lack of rigorous foundations for quantum field theory, these successes provide great encouragement to physicists that their ideas are fundamentally sound. (Atiyah 1995: 6070)

What does this enhancement of credibility involve?

Duality arguments, where a problem situation is transformed into a dual one which may be simpler to solve, crop up all over in mathematics, from projective geometry to Fourier transforms. In contemporary physics an important example concerns what are known as *mirror manifolds*. Here physical arguments are given to suggest that string theories in two manifolds are dual to each other, with the consequence that parameters governing the quantum geometry in one model are transformed into ones corresponding to classical features of another. In one case of this mirror duality the difficult task of counting the number of holomorphic curves on a Calabi–Yau manifold becomes in the dual mirror manifold the simpler question of how certain cohomology elements vary as its complex structure varies. This generated a range of mathematical predictions, later confirmed, of the number of curves of genus g and degree d on a certain quotient of a quintic hypersurface in complex projective 4-space.[4]

As one mathematical physicist puts it:

> One should note, however, that very rarely can one actually *prove* (even in the physics sense of this word) that two given physical systems are dual to one another. Often the existence of dualities between two given physical systems is guessed at based on some physical consistency arguments. Testing many non-trivial consequences of duality conjectures leads us to believe in their validity. In fact we have observed that duality occurs very generically, for reasons we do not fully understand. This lack of deep understanding of duality is not unrelated to the fact that it leads to solutions of otherwise very difficult problems. At the mathematical level, evidence for duality conjectures amounts to checking validity of proposed solutions to certain difficult mathematical problems. (Vafa 1998: 540)

The important things to note are that these consequences are purely mathematical and that their 'non-trivial' nature, implying a lack of likeliness, is being taken into account. The verification of these unlikely numerical predictions makes it almost certain that the corresponding physical systems are

---

[4] We are dealing with large numbers here, e.g., 317206375 curves of genus 0 and degree 3 (cf. Giventhal 1996). Numerical coincidence is out of the question. Let us note here for future reference that mirror symmetry is expressible as an equivalence of categories.

dual, thereby increasing confidence in the physical consistency arguments which predicted such duality. So, the confirmation of these mathematical predictions gives mathematical physicists the confidence that 'their ideas are fundamentally sound'. Are they wrong to react like this? We should demand of Bayesianism, or any other confirmation theory, that it be able to account for this practice, or else be in a position to discount it as irrational.

Through these case studies we have seen how thoroughly intertwined are factors of uncertainty in mathematics and physics. Although I recorded above Ian Stewart's comment that it is 'bizarre' to find experimental evidence bearing on a mathematical result, in a sense we could say that this happens all the time. What I have in mind here are computer calculations, where the behaviour of some carefully configured semiconductors is taken as having a bearing on a mathematical question. Perhaps this type of interrelation is overlooked owing to the flexibility of the computer, where it appears that it is the software that is doing the computational work, rather than the physical hardware, whereas Libchaber's liquid helium is directly involved in the production of a value for Feigenbaum's constant. An intermediate case is the use of reacting strands of DNA to allow certain computations to take place. As I mentioned in chapter 5, Fallis (1997) describes the employment of such biochemical computers to determine the existence of paths through specified graphs in the context of an argument to the effect that, as regards confidence in a hypothesis, there is no *qualitative* difference between, on the one hand, evidence gained by such means and, on the other, mathematicians' potentially fallible hand-written proofs. His arguments support my position that physical evidence impacts on degrees of belief in mathematical theories and vice versa. Let us now try to make more sense of how we might construe the types of confirmation we have just seen in terms of scientific models.

6.3 MATHEMATICS AND SCIENTIFIC MODELS

Bayesianism is usually represented in a formal fashion as an extension of the syntactic approach to the description of scientific theories, where evidence is held to support a theory if it is a deductive consequence of that theory. However, as the case studies outlined in section 6.2 demonstrate, scientific modelling, a subject currently receiving considerable philosophical attention, is a subtle affair. In *Models as Mediators* (Morgan and Morrison 1999), the editors advocate the philosophical study of modelling as an essential component of scientific practice:

The implication of our investigations is that models should no longer be treated as subordinate to theory and data in the production of knowledge. Models join with measuring instruments, experiments, theories and data as one of the essential ingredients in the practice of science. (Morrison and Morgan 1999: 36)

Contributors to the book maintain that models operate in a variety of ways to mediate between theories and experimental data, while retaining some considerable degree of autonomy. No longer can we see them in the Tarskian sense as set theoretic entities satisfying the axioms of some scientific theory, and there is far more at stake for models in the discovery and justification of scientific knowledge than merely belonging to the intended class of models interpreting some particular theory, as they claim the so-called semantic view of theories decrees.

On the other hand, advocates of this semantic view (e.g. da Costa and French 2000) see the *Models as Mediators* programme as overstating the autonomy of models from theory, and also as misrepresenting their position as one which makes the naïve claim that the production of models from theory is a kind of automatic process.

However this debate is settled, I think it is fair to say that the only hope for Bayesianism to prosper is for it to adopt a looser characterisation of scientific inference, in terms of modelling, than has typically been the case. Even in the case of mathematical inference, the logical representation of which is generally thought to be considerably less of a distortion, we have seen that a less formal language was necessary to express the prospects of success for proof strategies, the strength of analogies, vague connections between theories and the very varied types of mathematical evidence.

Let us consider the third case study from section 6.2. Equilibrium statistical mechanics (ESM) is a discipline which has provided a host of models to aid our understanding of phase transition phenomena in a variety of physical situations: spontaneous magnetisation in ferromagnets, oscillations in crystal lattices gases, the surface tension of a liquid, etc. Its techniques have also lent themselves to relativistic quantum field theory and differentiable dynamical systems. The traditional Bayesian approach might have talked about the change to one's degree of belief in ESM conditional on establishing observations of, say, the phenomenon of spontaneous magnetisation. However, this approach is faced with the problem of delineating ESM as a theory neatly expressible as an axiomatic theory. It is hard to conceive of the totality of the non-observational components of a theory such as ESM as the logical conjunction of a bundle of propositions. What, then, is a specified degree of belief in ESM supposed to indicate? I suggest that the best one can say is that it indicates confidence in ESM's capacity to generate good models of relevant phenomena – or, in Kuhnian terms, its capacity to solve relevant puzzles.

The two-dimensional Ising model is not deemed to be an accurate representation of a ferromagnetic sample. Much of what is known about interatomic forces is bracketed out and unrealistic assumptions are made to allow the construction of a simple and mathematically tractable member of a universality class of models (Hughes 1999), to which less crude models of various kinds of process in condensed matter physics are thought to belong. Many of the features of the Ising model will not be applicable to ferromagnets. What we find here is not a direct confirmation from evidence to model as an intended representative of a theory, but rather the 'adequate' treatment of a phenomenon. Practitioners of ESM have been able to garner the resources to manage to account in a reasonable way for the phenomenon of spontaneous magnetisation. To the extent that confirmation has taken place, we ought to be led to place greater confidence in the theory to do likewise in similar situations, that is, that the relevant techniques will cope well in those situations where the theory is deemed appropriate. Given ESM's success with crystalline structures, one might have expected it to be able to deal with glasses. However, it turns out that the microscopic structure of glass is not in equilibrium and so we find that:

> glasses are outside the piece of reality that is well described by equilibrium statistical mechanics. (Ruelle 1991: 189)

Failure on ESM's part to account for the properties of glass is, therefore, not to count against it. To sum up, in typical scientific situations, confirmation of a theory involves learning that upon it one may base a satisfactory account of a phenomenon understood to lie within its purview.

Now, before I set this view of confirmation the task of making sense of the old evidence problem, I want briefly to make the case for a deeper engagement of philosophy of science with contemporary mathematics. Let us then consider the fourth case study, which represents a different kind of confirmation. If physicists can be led to make mathematical predictions which would appear to be extremely unlikely to be correct without their arguments, then there is very likely something sound about the processes by which they are attained. For instance, the physical intuitions which suggest that two models are dual to one another must be deemed likely to yield further such dualities. We are not talking here about the correctness of a model, but rather of the conjectural interrelation between models. These may be 'toy' models which are investigated with no prospects anticipated for direct representation of some piece of reality, nor even of their belonging to a universality class, as with the Ising model. Rather, they are studied with a view to exploring a space of models, including, or at least juxtaposed to, one or more with the potential for eventual integration with observational data, such as when toy models of 2 + 1-dimensional quantum gravity are examined to see if three-dimensional general relativity emerges in the classical limit in the hope that models that do so will be useful for constructing good models of quantum gravity in 3 + 1 dimensions.

An attitude one finds among some philosophers of physics is to avoid consideration of this kind of speculative theoretical physics. This is understandable. In that philosophers have been guilty in the past of neglecting the subtle relationship between modelling and experimentation, the last thing we need look at, one might think, is theorists' highly abstracted mathematical pictures of the world. Would this not return us to the earlier unbalanced concentration on scientific theories? But to turn away from mathematical conceptualisation just as modelling becomes the focus of philosophical contemplation may be undesirable as it will prevent us from examining the richly structured 'trading zones', to use a phrase of Peter Galison (1997), currently emerging between physicists and mathematicians, and the mutual benefit accruing from their contact. Curiously, in the collection referred to above (Morgan and Morrison 1999), it is a philosopher of economics, Marcel Boumans (1999), who is most aware of the conceptual contribution of mathematics to the production and establishment of models. Boumans portrays several ingredients being integrated into a model – theoretical notions, mathematical concepts, mathematical techniques, stylised facts, empirical data, policy views, analogies and metaphors – in such a way that if its construction satisfies certain criteria then it is accompanied by a certain built-in justification.[5] Think, then, how much more the role of mathematics needs to be stressed in a discipline such as physics where the cross-pollination of ideas with mathematics is so fertile.[6]

---

[5] Chitra Ramalingam, while an MPhil. student in the History and Philosophy of Science Department at Cambridge University, has further observed in an essay that where Boumans talks about mathematical arguments using the past tense, suggesting a interest in them as historically situated, the other contributors describe such arguments using the 'disembodied present tense'.

[6] One only has to think of the impact knot theory, quantum group theory, 2D integrable models and quantum field theory have been having on one another since the mid-1980s. Here, one could almost conceive of a type of confirmation where the scientific theory generates 'important', rather than correct, mathematics. Incidentally, in the essay mentioned in n.5, Ramalingam shows the inaccuracy of Norton Wise's construal of Kelvin's working out of the analogies between gravitational potential, water flow, heat flow, and electrostatic distribution, solely via physical interpretations. It was, in fact, at first mathematics-led. While Cartwright who 'learned about Kelvin from Norton Wise' (1999a: 48n)

Philosophers of science seem to be all too ready to dismiss the creative contribution of mathematics to scientic thinking. Where Wigner found it mysterious that mathematics developed for aesthetic reasons should find uses in the physical sciences, this is typically deflated by the twin strategies of taking mathematics to be a transparent language which plays no role in shaping the way we think about the field it is used to describe, and then of pointing out how so much of mathematics arises from scientific theorising. Mathematicians and scientists know that this is wrong. Indeed, Ruelle (1988) clearly sees himself as adopting a controversial position for arguing for the existence of even a small amount of mathematics reached only via physical considerations. Mark Steiner (1998) has pointed to the Wignerian faith physicists display in assuming that mathematical arguments will prove to have physical significance. Unfortunately, his radical conclusion that this faith is justified by our inhabiting a 'user friendly universe' may well lead people to overlook the subtleties of the changing relationship between mathematics and science.

In a recent article, Lawrence Sklar (2000) has called on philosophers to carry out a thorough study of the aims and methods of *interpretation* as a scientific activity. Rather than seeing interpretation as mere philosophical gloss, Sklar views it as an essential part of justificatory activity, since

> time and again, what we find is that foundational theories, for all their empirical success and all the good reasons we have for thinking of them as at least 'pointing toward truth', if not finally giving it, exist themselves in a perpetual state of indeterminacy and conceptual lack of 'fixedness'. They come replete with internal puzzles that leave us perplexed, at the same time that we are assured that they are indeed on the right scientific track, about just what the theories are saying about the world and what kind of a world they can be describing. (Sklar 2000: 735–6)

Here again, as with those studying model building, we find a scientifically knowledgable philosopher aware of the distortion of oversimplistic representations of the relationship between evidence and theory. Sklar's portrayal is one that suits me well, since mathematicians can and do contribute to the interpretation of physical theories through their deeper understanding of the mathematics involved in them. It seems not to be widely recognised that mathematics itself is also an open-ended interpretative discipline, contemporary philosophy of mathematics having done much to obscure this fact. To arrive at a more adequate view will require a revolution in our

---

can distance herself from 'philosophers studying the mathematical structures of our most modern theories in physics' (*ibid.*: 4–5), all seem to suffer from a certain blind spot when it comes to the contribution of mathematics.

outlook to stop taking mathematics for granted. Here we turn to the more modest goal of examining how the conceptions of confirmation presented in this section fare with the thorny problem of old evidence.

6.4 THE PROBLEM OF OLD EVIDENCE

One of the first serious obstacles to a Bayesian rendition of scientific inference, as presented by Earman (1992), is the problem of old evidence. Put simply, according to Bayes' theorem, if we know a piece of evidence to be true, a newly devised theory which accounts for that evidence cannot have its credibility boosted by the data, even if there is no other theory around which can account for it. In view of the liberality of Bayesianism as a way of representing reasoning, too generous some would argue (Albert 2001), to think that it cannot cope with as a common a scenario as old evidence strains belief. It is as though Borges had told us of a country where a style of prose had been devised which no piece of paper in that realm would allow to have written upon itself. Let us first see how old evidence operates in mathematics with an example of Pólya's.

Pólya (1954b: 5–6) explains how when you are considering the plausibility of a conjecture, you look to see if it works in specific cases that you know already. Adapting his simple example slightly, you are set the task of finding a formula for the curved surface area of a frustrum of a cone (imagine here a cone with its nose chopped off). With your shaky and rudimentary calculus you arrive at the conjecture that the surface area is $\pi (R + r) \sqrt{\{(R - r)^2 + h^2\}}$. You are moderately confident that this is correct, but you accept that you may have erred. Later it occurs to you that you know the answer in certain degenerate cases: the area of an annulus is $\pi (R^2 - r^2)$, the curved surface area of a cylinder is $2\pi rh$ and the curved surface area of a cone is $\pi R \sqrt{(R^2 + h^2)}$. It is also clear that inverting the frustrum should not alter the area. You check your formula's accuracy for $h = 0$, $R = r$, and $r = 0$, and symmetry on exchanging r for R. By now your confidence has grown to near certainty. What could be unreasonable about this boost in confidence on discovering that a general proposition has already known true consequences? It could be objected that one should have already known that these were consequences and so have factored this knowledge into one's belief of the general proposition, but we dispensed with that idea in chapter 5.

The classic example of old evidence providing a boost to a novel theory is the capacity of the General Theory of Relativity (GTR) to account for the anomalous advance of Mercury's perihelion (AMP), known for several years prior to 1915 to be roughly 43 arcseconds per century. Bayes's theorem gives us

$$\Pr(\text{GTR}|\text{AMP}) \;=\; \Pr(\text{AMP}|\text{GTR})\cdot\Pr(\text{GTR})/\Pr(\text{AMP})$$
$$\;=\; \Pr(\text{GTR}),$$

since, by 1915, $\Pr(\text{AMP}) = 1$ and, as AMP is a consequence of GTR, $\Pr(\text{AMP}|\text{GTR}) = 1$. Thus, on the face of it, Bayesianism would appear to offer as methodological advice to the scientist not to bother formulating a theory which accounts for already known data, a bizarre piece of counsel, and yet many scientists found this success of GTR's very convincing.

Naturally, much effort has been invested in rectifying this difficulty. The formula suggests that the only ways to do so are to allow the scientist hearing of Einstein's ideas the possibility that $\Pr(\text{AMP}) \neq 1$ or to abandon $\Pr(\text{GTR}|\text{AMP})$ in favour of some other conditional.[7] These possibilities relate to the two general approaches to resolving the paradox, where we must contrast an agent's current belief with an imagined situation:
1. Imagine that the agent does not possess the experimental data:
    (a) Imagine an agent's history had been such that she never came to believe the evidence
    (b) Imagine that the evidence, or the observations on which the evidence is based, has been deleted from the agent's background knowledge.
2. Imagine that the agent does not believe that the theory accounts for the data:
    (a) Imagine that the agent does not know that the evidence is a consequence of the theory – i.e. allow for conditionalising on 'logico-mathematical learning', and consider probabilities of the kind $\Pr(\text{GTR}/\text{GTR} \vdash \text{AMP})$
    (b) Imagine that the observed phenomenon is known to have been brought about by a minimal piece of divine interference.[8]

Objections can be raised to each of these approaches. For example, in the case of 1(a) we must wonder what to do if the observation played some role in the production of the theory, as was the case in the example we are considering. Einstein is known to have rejected other candidate theories because they produced incorrect perihelion advances. With 1(b), a common concern is that neat excision of a statement from one's background knowledge will not be possible.

---

[7] Alternatively, we adopt a method of updating other than conditioning, such as Jeffrey's *reparation* (Jeffrey 1992: 103–7).
[8] See Barnes (1999) for the second part of 1(b) and 2(b).

As for 2(a), it would seem to fit the bill with Pólya's example. However, in the case of Mercury, Earman protests that it is not learning that AMP is a consequence of GTR that is at issue but the confirmatory power of the astronomical evidence of Mercury's behaviour. Painted in these stark terms one will incline to agree with Earman. The squiggles on a piece of paper which represent the derivation do not seem enough when Mercury is known to be out there misbehaving. But was this the only choice? Let us consider this point from the other direction. Surely it is problematic to say that the observational evidence of Mercury's behaviour confirmed Einstein's theory before he devised it. But what is gained by the alternative of saying it becomes evidence as soon as Einstein utters the theory, but before he has made the link between theory and observation? Would it not be better to have both the acceptance of the data and the derivation as components of the larger process of accepting that the theory accounts well for a phenomenon within its domain? Scientific observations become evidence when they are taken up within the ambit of an explanation.

Barnes' depiction of divine intervention 2(b) would seem to be closely related to this line of thought. We are to contrast our confidence in a theory after discovering that it can account satisfactorily for a phenomenon understood to be within its purview, with the level of confidence in the theory were we to discover that the phenomenon was a product of some exogenous process, e.g., it had been brought about by divine fiat, and so no longer to be taken to be within the theory's range. It is interesting to note that in devising 2(b) Barnes has hit upon a solution which involves an idea recently developed by Judea Pearl (2000). What it concerns is the difference between observing the value of an operator and exogenously fixing the value of that operator. Pearl has shown how important it is to expand the probability calculus to allow entities such as $\Pr(Y=y|\mathrm{do}(x))$, where the value of the variable X is forced to be x. The problem for Barnes' approach, as he points out, arises when the act of fixing the evidence causally brings about the occurrence of the hypothesis, as when observing that someone smokes forty cigarettes a day and forcing them to do so both bring about lung cancer.[9]

Let us continue with Earman's doubts about the Bayesian handling of old evidence. He gives a list of four factors which might be taken into account by anyone explaining why the perihelion data provided good confirmation of General Relativity: precision; no adjustable parameters; a good bootstrap test; and, dozens of attempts had already failed. His worry is that:

[9] Pearl goes to great lengths to develop a calculus for the 'do' operator to cope with such causal situations.

without solving the Bayesian problem of old evidence, we can recognize on independent grounds the confirmational virtues of the perihelion data. Of course,
there is nothing to block Bayesians from taking into account the factors enumer-
ated above. But how these factors can be made part of Bayesian calculations in the
context of old evidence remains to be seen. (Earman 1992: 132)

But the occurrence of the first three factors is what constitutes GTR as
accounting *well* for AMP, while the fourth bears on the prior probability
of GTR accounting for AMP.[10] By 1915, it seemed unlikely that it would be
possible for Newtonian mechanics to be able to account for the perihelion
data without making some dubious assumptions. So, `Pr(GTR)`, as short-
hand for `Pr(GTR can account well for all physical phenomena within its
scope)`, will start out low as it is realised that it is no simple matter to be
able to cope with all relevant cosmological phenomena, especially in view
of the failure of so many of its predecessors. AMP had plagued cosmology
for several years, provoking the postulation of all manner of ingenious hypotheses, such as the presence of intramercurial matter of various forms.
At this point `Pr(GTR/GTR accounts well for AMP)` would surely be
higher than `Pr(GTR)`. Historically, many scientists heard of GTR and
its success with AMP simultaneously, in which case this success was factored directly into a prior `Pr(GTR)`. For others,[11] a positive difference
would have existed between `Pr(GTR/GTR accounts well for AMP)`
and `Pr(GTR)`.

The elements of an 'accounting for' involved in the Einsteinian account
of Mercury's perihelion advance include: the general laws of GTR, approximate modelling assumptions, results in differential geometry, and observational evidence, including theories of instrumentation. When we realise
this, we can see how uncertain mathematical knowledge may need to be
factored in. Perhaps the AMP case is not well suited to allow us to see what
is going on, since the mathematics was readily available, but we have already
discussed a case where the proof of a mathematical result gave greater
credence to a scientific theory, namely, the Lee–Yang circle theorem. The
boost given to equilibrium statistical mechanics and in particular to its relevance to ferromagnetic spin systems by the confirmation of a mathematical
prediction may act as our guide here. ESM's ability to account for spontaneous magnetisation in ferromagnets was a good test for it. We would have

---

[10] Okasha (2000) also argues for explanatory considerations to be construed in a Bayesian light. Might
an attempt to model the notion of accounting well for phenomena, involving prior degrees of belief
that a theory could do at least as well as it turns out to have done, introduce a quantitative element
into this kind of Bayesianism?

[11] If nobody is in this class, a small thought experiment would allow the AMP postdiction to take place
sufficiently later then the publication of GTR.

`Pr(ESM can account for phenomena within its range/ESM can account for spontaneous magnetisation) > Pr(ESM can account for phenomena within its range)`. After the modelling, but prior to the proof of the theorem, to calculate `Pr(ESM can account for spontaneous magnetisation)`, we would have required `Pr(Lee—Yang circle theorem)`.

On the other hand, had the circle theorem been already known, as it easily might have been, the same kind of boost to ESM would have occurred as the ingredients (modelling, mathematical theorem, calculations, assumptions, observations) came together in a different order. 'Accountings for' are always new; there is no old evidence. If observations have not been linked by a model to a theory, they may be old but they do not provide evidence.

6.5 CONCLUSION

We should recognise that within the philosophy of science Bayesianism, as a theory of confirmation, is not alone in its neglect of plausible mathematical reasoning. Any philosopher of science must either deny the epistemic role accorded by contemporary scientists to the confirmation of mathematical predictions arising from their research, or else account for this practice within their theory of scientific rationality. I have argued that the Bayesian philosopher of science who does not see fit to contradict the reasoning of leading mathematical physicists should consider degrees of belief in mathematical statements. Non-Bayesians in a similar position will have to find their own way out of this problem.

Thinking about evidence within mathematics makes us see how unhelpful it is to construe as timeless the evidential relation between facts and theories. Focusing on the mathematical component of scientific reasoning then makes plain that it is the achievement of a satisfactory explanation of a phenomenon which is at stake in confirmation.

PART III

The growth of mathematics

Within the philosophy of science, Bayesianism can be seen as positioned on the boundary between the logical empiricist legacy and a more practice-oriented, historical approach. Bayesianism in science points rather to the logical empiricist side, especially when it sees itself as an extension of logic. In mathematics, on the other hand, Bayesianism appears to point us away from logicism to the practice of mathematics. This we saw in the work of Pólya, someone very interested in what goes on behind the scenes. Now it was Pólya who suggested to his fellow Hungarian Imre Lakatos that he explore the early development of algebraic topology following on from the appearance of the Euler conjecture. Lakatos took him up on this advice, producing *Proofs and Refutations* – a dialectical account of the growth of mathematical knowledge.

Lakatos also came under the influence of Popper, whose negative attitude towards inductivism he shared. He could thus say of Pólya:

We owe this revival of mathematical heuristic in this century to Pólya. His stress on the similarities between scientific and mathematical heuristic is one of the main features of his admirable work. What may be considered his only weakness is connected with this strength: he never questioned that science is inductive, and because of his correct vision of deep analogy between scientific and mathematical heuristic he was led to think that mathematics is also inductive. (Lakatos 1976: 74n.)

We have already encountered this hostile attitude towards inductivism in science in chapter 2. Lakatos thought that the adoption of an inductivist logic such as Carnap's was misguided as it presupposed 'a fixed theoretical framework' (1968: 161), and warned that the 'inductivist delusion', which imagined that 'truth (or some quasi-truth like probability)' (1978b: 41–2) could be transmitted upwards from basic statements to axioms, would incline practitioners to remain within that framework. Both deductivism and inductivism, therefore:

trade the challenge and adventure of working in the atmosphere of permanent criticism of quasi-empirical theories for the torpor and sloth of a Euclidean or inductivist theory, where axioms are more or less established, where criticism and rival theories are discouraged. (Lakatos 1978b: 42)

In response, one could say that, following Kuhn, mathematics needs periods of 'normal mathematics' as much as science needs 'normal science'. Perhaps we could then adopt the position that Bayesianism may be insightful during periods of relative conceptual stability. This would make sense of an impression many have had that Pólya was more a great puzzle solver than a major innovator. Hermann Weyl, for instance, once wrote of him in a reference:

His way of doing mathematics is really completely foreign to me. He is to a lesser degree concerned with knowledge but rather with the joy of the hunt. However, I admire his brilliance extraordinarily. His ideas are certainly not of the type that would cast light on the major relationships of knowledge. His papers are rather single, bold advances toward very specific, limited points in an undiscovered land that will remain totally in the dark. (Alexanderson 2000: 38–40)

In chapters 7–9 we turn to consider the growth of mathematical knowledge. In chapter 7, we examine Lakatos's writings, in particular *Proofs and Refutations*. Hopelessly inaccurate even as a schematic history of early algebraic topology, somewhat superficial as a phenomenology of discovery, too captivated by its struggle with an imaginary opponent, the formalist, to make more of the excellent choice of its subject matter, this work still remains an oasis in a desert of neo-logicism. In chapter 8, we explore the prospects facing Lakatos had he lived long enough to transfer his notion of research programmes back from science to mathematics. Finally, in chapter 9 we examine a controversial case of conceptual development to discover the kinds of quality mathematicians expect from novel theories.

CHAPTER 7

Lakatos's philosophy of mathematics

7.1 INTRODUCTION

Nearly forty years have passed since Imre Lakatos published his paper *Proofs and Refutations* over four successive issues of the *British Journal for the Philosophy of Science*. Two years after his untimely death in 1974, these articles, along with other sections of his doctoral thesis, were published as a book (Lakatos 1976). It was only then that (generally favourable) reviews of his work appeared, among them notably one by Quine (1977). Uncharacteristically for a piece of philosophy of mathematics written in the latter half of the twentieth century, even mathematicians appeared to enjoy it. Davis and Hersh (1981: 347), themselves mathematicians, maintain that before it appeared in book form it was a 'sort of underground classic among mathematicians, known only to those intrepid souls who ventured into the bound volumes of the *British Journal for the Philosophy of Science*'. And yet, despite receiving praise from mathematicians and positive reviews from philosophers, very few attempts have been made to build on Lakatos's ideas.

I shall begin by assessing the later Lakatos's portrayal of the way mathematical theories develop. Then, after defending him from a number of criticisms that have been aimed at his work, I shall be arguing later in this chapter that Lakatos was overly concerned with questioning the status of the accepted statements of an established mathematical theory, and that accordingly he did not pay sufficient attention to its more conceptual features, including its relationships with other theories, an imbalance which was not so noticeable in his treatment of informal theories, but which emerges quite plainly when he touches on modern mathematics. By confining uncertainty in mature theories to the issue of whether the full content of informal prior theories had been captured, while recognising the achievement of a kind of stability in rigour, he was led to the mistaken position that the development of modern mathematics lacks much of the freedom and excitement of earlier times. This constitutes a serious misunderstanding on Lakatos's part and prevented him from observing an important part of the dialectical process involved in the development of the mathematics of the twentieth century. I shall suggest instead that the appropriate use of rigorous definition and axiomatisation has not acted as a hobble on the creativity of mathematicians, but rather an invaluable tool in the forging of new mathematical theories and the extension of old ones.

Although I shall be concentrating in this chapter on the limitations of Lakatos's philosophy of mathematics, I hope my appreciation of his work will be evident. One does not choose to criticise and develop a philosophical position unless one takes it seriously. Not only has Lakatos handed down to us important notions such as that of a 'proof-generated concept', but through his case studies he has let us hear for the first time a snatch of mathematical conversation. What has perhaps appealed most to mathematicians about *Proofs and Refutations* is the sense of authenticity Lakatos managed to convey by casting the book in the form of a dialogue in which the characters are frequently made to express themselves in the words of various prominent nineteenth-century mathematicians. Yet while Lakatos has rightly become the patron saint of those struggling to produce an adequate philosophy of mathematics, this should not blind us to his faults. Either one takes the text of *Proofs and Refutations* as providing a 'rational reconstruction' of the early history of algebraic topology, while the footnotes point to discrepancies existing between this version and the often irrational course of real history. But then as Koetsier (1991: sec. 1.3.3) has observed, even the material in the footnotes is seriously historically inaccurate. Or else, like Glas (1993), one chooses to ignore Lakatos's introductory claim that 'the real history will chime in in the footnotes' (Lakatos 1976: 5) and side-steps Koetsier's criticism by taking the text and footnotes as a whole to be a rational reconstruction. Either way, however, as we shall see, Lakatos omitted to tell us about the very heart of nineteenth-century algebraic topology in his endeavour to shape his account into the method of proofs and refutations, ignoring for the most part the rationale for Poincaré's work on the Euler conjecture.

7.2 LAKATOS'S STAGES OF DEVELOPMENT

The final glimpse we have of Lakatos's philosophy of mathematics is in the second section of *The Method of Analysis-Synthesis* (1978b: 93–103) which is based on an address he gave in 1973, the year before his death. In this paper Lakatos portrays the development of a mathematical theory according to the following schema. Mathematical theories pass through three phases in their development: the first stage of discovery is the stage of 'naive trial and error' in which the 'naive conjecture' is reached by the method of 'Popperian conjectures and refutations'. In the case study described in *Proofs and Refutations*, Lakatos claims, this stage lasted 2,000 years from the time of Euclid to Descartes.[1] The second stage is that of 'proof-procedure' in which, via the *method of analysis and synthesis*, an expression used interchangeably with the *method of proofs and refutations*, 'the naive conjecture disappears, the proof-generated theorems become ever more complex and the centre of the stage is occupied by the newly invented lemmas, first as hidden (enthymemes), and later as increasingly well articulated auxiliary assumptions. It is these hidden lemmas which, finally, become the hard core of the programme' (1978b: 96). The third stage is then that of the *research programme*, a concept he had invented for his philosophy of science. We may suppose then that had Lakatos returned to the philosophy of mathematics, he would have elaborated on this third stage. However, as we shall see in chapter 8, the ambivalence he displays in his attitude towards axiomatic mathematical theories would have made this problematic for him.

Lakatos makes it clear that he sees these three stages in the life of a theory as distinct in character. The method of proofs and refutations, or of analysis and synthesis, unavailable during the first stage, is the dominant dynamic of the second stage and plays only a subsidiary role in the third stage. Thus:

in the creative development of algebraic topology we rarely any longer find analyses. Once the lemmas become corroborated and even organised in axiomatic systems, once the mathematical machinery is established, analysis, 'working backwards' may still be applied as a heuristic tool in puzzle-solving, but it becomes clear that its role is only psychological. It helps the imagination to produce valid proofs or explanations in terms of a *given* research programme. Analysis in mature science and mathematics no longer leads to revolutionary progress. Analysis is only revolutionary when it engineers a breakthrough from a low-level naive conjecture to a research programme. (1978b: 99)

---

[1] It is very easy to misread the paragraph on p. 96 in which he discusses this stage as identifying the Popperian method with the method of proofs and refutations. The editors falsely remark in a footnote that this stage is described in chapters 1 and 2 of *Proofs and Refutations*, which comprise the bulk of the book. However, a close reading of this paragraph in conjunction with pp. 6 and 7 (especially the footnotes) of *Proofs and Refutations* shows that Lakatos intends us to think of the first stage as the production of a formula that works for a range of common polyhedra. Indeed, he tells us of Euler, prior to any proof, testing the formula against various prisms. We have seen in chapter 4 that Pólya, who suggested this case study to Lakatos, used this process of guessing and testing as an example of inductive reasoning in mathematics.

Once the excitement of the second stage is over we enter the more staid third stage where 'imagination is tied down to a poor recursive set of axioms and some scanty rules' (1978b: 68).[2]

This picture of mathematical development will provide us with a framework in which to locate possible criticisms of Lakatos's ideas. The two critics I shall focus most closely on here are Mark Steiner (1983) and Solomon Feferman (1981). The latter professes a certain sympathy with Lakatos's cause:

> Personally, I have found much to agree with both in his general approach and in his detailed analysis. (Feferman 1981: 310)

However, Lakatos's account he finds 'too single-minded and much more limited than he makes out' in that 'he plays only a single tune on a single instrument – admittedly with a number of satisfying variations – where what is wanted is much greater melodic variety and the resources of a symphonic orchestra' (*ibid*.: 310). Steiner, on the other hand, is generally hostile, reading Lakatos as holding that we can have no mathematical knowledge, a view he naturally wishes to discredit having expressed contrary opinions in a book entitled *Mathematical Knowledge* (Steiner 1975).

I shall be using the history of algebraic topology to provide illustrations both of what is missing and what is incorrect in Lakatos's account and so shall require an idea of the possible timings of the transitions between the stages. Lakatos himself (1976: 6n.) dates the transition from the first to the second stage of discovery in algebraic topology to the discovery by Euler of the '$V - E + F = 2$' formula published in 1758 (or possibly to a manuscript of Descartes *c.* 1639). However, neither date sits well with the claim that the method of proofs and refutations was discovered only in 1847 by P. L. Seidel (*ibid*.: 136).

Lakatos's rational reconstruction does not extend to the onset of the final stage of development of this branch, that is, the appearance of the further 'derivatives' of the hidden lemmas found in the second stage as the axioms of algebraic topology. However, on the basis of comments he makes elsewhere

---

[2] The account Lakatos presents here runs against an indication he gives elsewhere that *Proofs and Refutations* describes a research programme (Lakatos 1978a: 52n.). Glas (1993) rightly underlines the similarities between the methodology of proofs and refutations and the methodology of research programmes and would have the 'idea that the relationship $V - E + F = 2$ . . . expresses some fundamental feature' belong to the hard core (1993: 49). This I would agree is the correct way to proceed and below I sketch the hard core of modern algebraic topology in terms of an aim, but it contradicts Lakatos who resorted instead to hidden lemmas become axioms as the constituents of the hard core (1978b: 96).

(Lakatos 1978b: 66), it seems likely that he is referring to Eilenberg and Steenrod's axiomatisation of homology and cohomology theories.[3] On the other hand, algebraic topology was on a fairly firm footing by around 1920 after the application of Brouwer's innovations to Poincaré's less rigorous work by the Princeton topologist, J. W. Alexander. Indeed, Dieudonné (1989: 16) dates the acquisition of rigour in this branch to Brouwer's work of 1910.

I raise the question of Lakatos's dating of this transition for the reason that the later this is fixed the less credible the idea that the method of proofs and refutations suffices to account for growth during the second stage, while the suggestion that the imagination of algebraic topologists was 'tied down to a poor recursive set of axioms and some scanty rules' already incorrect for the later date becomes patently absurd for the earlier one when we come to consider the significant contributions made by topologists such as Lefschetz, Cech, Vietoris and Hopf in the 1920s and 1930s.

7.3 THE METHOD OF PROOFS AND REFUTATIONS

I shall keep my comments about the first stage and the transition to the second stage brief. Feferman wonders where the initial proof originated from and what happened before the onset of the method of proofs and refutations (criticisms vi and i).[4] It is true we get to hear very little about what happened before Cauchy's proof, which Feferman describes as already well advanced, other than what is mentioned in two footnotes (Lakatos 1976: 6–7), but Lakatos does say there that his discussion starts where Pólya's stops. It would of course be desirable to account for the evolution of theories prior to the 1840s, yet had he succeeded in the restricted project of describing mathematical change since that time, then it would surely still have been a magnificent achievement.

Now let us turn to the second stage, in which the *method of proofs and refutations* prevails. Lakatos describes the seven stages of the growth of informal mathematical theories via this method as follows:

---

[3] Eilenberg and Steenrod (1952), the results of which had been announced seven years earlier. Leaving aside problems with the idea of characterising a branch of mathematics in terms of its axioms, which I shall discuss later, it is worth briefly remarking that by 1952 homotopy theory had become just as important a part of algebraic topology as homology and cohomology theory, marking its difference by its failure to satisfy Eilenberg and Steenrod's 'Excision axiom'. In their book (1952: 49) Eilenberg and Steenrod date the axiomatisation of the homotopy groups to a paper of J. Milnor later published in 1956.

[4] I shall follow Feferman's enumeration of his questions and criticisms from (i) to (x) (1981: 316–20).

(1) Primitive conjecture.
(2) Proof (a rough thought-experiment or argument, decomposing the primitive conjecture into subconjectures or lemmas).
(3) 'Global' counterexamples (counterexamples to the primitive conjecture) emerge.
(4) Proof re-examined: the 'guilty lemma' to which the global counterexample is a 'local' counterexample is spotted. This guilty lemma may have previously remained 'hidden' or may have been misidentified. Now it is made explicit, and built into the primitive conjecture as a condition. The theorem – the improved conjecture – supersedes the primitive conjecture with the new proof-generated concept as its paramount new feature.
(5) Proofs of other theorems are examined to see if the newly found lemma or the new proof-generated concept occurs in them: this concept may be found lying at the cross-roads of different proofs, and thus emerge as of basic importance.
(6) The hitherto accepted consequences of the original and now refuted conjecture are checked.
(7) Counterexamples are turned into new examples – new fields of inquiry open up.

(Lakatos 1976: 127–8)

Feferman's criticisms of the proof-procedure stage amount to the claims that: (vii) Lakatos has treated only proofs of conjectures of a particular form, namely, $\forall x[A(x) \rightarrow B(x)]$; (viii) logical analysis can account for these cases just as well; and (iii) counterexamples do not form the principal feature of the (self-)critical examination of proofs. To the first point one might say that this format is relatively common in the sort of conjecture that spearheads the drive towards the establishment of a research programme. Leafing through Hilbert's twenty-three problems one finds that many of the specific problems to be resolved are of this type. The second point is one with which Lakatos would not have disagreed strongly. He himself made a rather unsuccessful attempt to formulate a symbolic scheme of the method of proofs and refutations in *The Method of Analysis-Synthesis* (Lakatos 1978b: 70) which the editors correctly point out does not agree with his verbal account. That Lakatos could attempt such a thing reveals a tension between, on the one hand, his appreciation of informal mathematics and, on the other, his preference for the logical structure of a proof over its more conceptual aspects. The last point, especially when conjoined to the claim that in any case proof analysis is by no means the only dynamic at play during the second stage, I would agree is on target. Here we can only say in Lakatos's defence that as his was the first attempt to characterise the development of informal mathematics, it was liable to be incomplete.

The false impression Lakatos gives is of a theory developing along a single track, driven by successive attempts to modify a particular conjecture by incorporating conditions brought to light by counter-examples to it. In chapter 1 of *Proofs and Refutations* he describes the development of work on the Euler formula up to the 1860s, then goes on in chapter 2 to discuss Poincaré's proof of 1899 which appears there out of the blue.[5] It is not pointed out, however, that the developments described in the earlier chapter formed only one, and by no means the most important, source for this later proof. Nor is it mentioned there that Poincaré considered his duality theorem, 'le théorème fondamentale', a far more important result.

To see an emerging branch of mathematics as crystallising around the seeds of a few scattered conjectures is surely inaccurate. It would be quite impossible to explain the appearance of Poincaré's work in algebraic topology, or *Analysis Situs* as he called it, contained in six papers running from 1895 to 1904, without describing the motivation that lay behind them. Fortunately for us, Poincaré himself recorded this for posterity in the summary he gave of his work, *Analyse de ses travaux scientifique*, written in 1901:

All the different paths along which I set out in turn led me to *Analysis Situs*. I needed the results of this Science to pursue my study of curves defined by differential equations and to extend this to higher order differential equations, in particular, to those involved in the three body problem. I needed them to study nonuniform functions of two variables. I needed them to study multiple integrals and to apply these to the expansion of the perturbation function. Finally, I glimpsed in *Analysis Situs* a way to approach an important problem in group theory, the study of discrete or finite groups contained in a given continuous group. (Poincaré 1921: 323, my translation)

As can be seen from this quotation, Poincaré studied algebraic topology to provide himself with tools for his other mathematical interests rather than from a desire to prove any particular conjecture.

While there are indications that Poincaré's appreciation of the work of Picard in algebraic geometry strongly influenced the *Analysis Situs* papers, which is reflected in the fact that the main applications of algebraic topology to other areas of mathematics were initially in this subject, the principal source for many of the ideas in these papers comes from Riemann's work in the theory of complex functions, most notably the study of integrals on Riemann surfaces.

---

[5] This is not quite correct. The proof given in *Proofs and Refutations* involves choosing coefficients for the chains from the field of integers mod 2, a technique introduced by Tietze only in 1908.

In both the introduction to the first of the *Analysis Situs* papers (Poincaré 1895) and the summary of his life's work mentioned above, it is Riemann and Betti that he considers as his predecessors and he makes no reference to the work on the Euler conjecture described in the first chapter of *Proofs and Refutations*:

In spite of everything, this branch of Science had until then been little cultivated. After Riemann came Betti, who introduced several fundamental ideas, but nobody followed Betti. (Poincaré 1921: 323, my translation)

I shall not add to what I said about Riemann's treatment of multiple-valued complex functions in chapter 4. An account of this episode can be found in Morris Kline's history of mathematics (Kline 1972: 655–65) from which it is clear that the method of proofs and refutations could not suffice as a historiographical tool. It might be argued that complex analysis had by this time reached the third stage of discovery, but this would point only to the lack of attention Lakatos gave to the influences which theories, possibly in different stages of development, have on each other.

To test the validity of Lakatos's ideas in a situation where they are most likely to succeed, I shall now examine an episode of proof-analysis which follows the method of proofs and refutations in broad outline, but which shows in its details the need for a richer theory. Riemann (for two dimensions) and Betti (for higher dimensions) had been interested in the situation where a collection of n subvarieties of a manifold U, all of the same dimension d, was such that the addition of an $(n + 1)$th variety of the same dimension would cause there to be a *homology* between them. That is, this last patch of space would make it possible for a variety of dimension $d + 1$ to be found whose boundary was formed by these $n + 1$ subvarieties. In this case the dth-dimensional Betti number of U was defined as $P_d = n + 1$. Poincaré took up this idea and tried to prove a duality theorem for manifolds which states that the Betti numbers of U of complementary dimensions p and $(h - p)$, where h is the dimension of U, are equal. However, without his being aware, he had made a subtle alteration in the definition of Betti numbers which went unrecognised until Heegaard, in a review of *Analysis Situs*, reported a supposed counter-example to the duality theorem. When Poincaré came to analyse what had gone wrong, he found that his definition of the independence of a collection of varieties, $v_1, v_2, \ldots, v_n$, differed from Betti's. For Poincaré such a collection is linearly dependent if there exists integer coefficients, $k_1, k_2, \ldots, k_n$, not all zero, such that the sum of the v_i taken with multiplicity k_i forms a homology, $k_1v_1 + k_2v_2 + \cdots + k_nv_n \sim 0$.[6] In *his* definition Betti had only allowed these coefficients to be 0 or 1.

By treating homologies like equations which could be added, subtracted, multiplied by an integer and divided by an integer if the n coefficients were multiples of that integer, the Betti numbers so produced may turn out to differ from those arising from Betti's original definition. The simplest example which demonstrates the difference between the two definitions is given by real projective space, RP(2), the space of lines through the origin in three-dimensional Euclidean space, in which there is a closed path which is not the boundary of any two dimensional subspace of RP(2), yet two copies of this path do form such a boundary. Think of rotating a plane passing through the origin by a half turn about a perpendicular axis. Then a line in this plane will be taken to itself, and yet the path of this line does not surround any region of RP(2). On the other hand, the path corresponding to a full turn of the plane bounds the space of all those lines not lying in the plane, essentially just as the equator of a sphere bounds its northern hemisphere.

Poincaré was concerned to find that his proof of the duality theorem seemed to apply equally well to both definitions. Although with his version of the Betti number Heegaard's space was no longer a counter-example to the duality theorem, it did contradict a lemma Poincaré had used in the proof of that theorem. The heart of the problem lay in the fact that Poincaré's definition of a subvariety allowed too much freedom to the way subvarities could intersect each other. Because of this a crucial lemma, which stated that for any collection of linearly independent varieties of dimension $(h - p)$, a p-dimensional variety could be found which intersected just one of these varieties and only once, was false. Poincaré had to redesign his proof and when he did so in the year after Heegaard's paper had appeared, it was by considering spaces as polyhedra, that is, as being constructed from varieties of all dimensions up to h intersecting in a controlled fashion. To prove the duality theorem under this scheme he defined for each polyhedron P a reciprocal polyhedron P' (polyèdre réciproque) and demonstrated with the use of linear algebra the following equalities between their Betti numbers: $P_d = P'_d$ and $P_d = P'_{h-d}$. From these equations the theorem then follows easily.

---

[6] The concept of a multiple of a variety was left vague. It was to be thought of as a set of varieties slightly deformed from each other.

If we compare this episode with Lakatos's schema for the Method of Proofs and Refutations shown at the beginning of this section, we find certain differences:

(1) The situation here involves the emergence of a counter-example, after the first proof of the theorem, which was not global but local in that it contradicted a lemma but not the theorem (stage (3)). Furthermore, the lemma had already been made explicit.

(2) It would be a mistake to describe Poincaré's subsequent proof as resulting from lemma-incorporation (stage (4)). Although the two proofs are conceptually similar, the second differs considerably from the first and it would not be possible to make a comparison using a simple predicate symbol schema. Thus, while the strategies behind both proofs are comparable – first characterise intersections between varieties of complementary dimensions and then compare their linear dependencies – and while one can see how problems with the first strategy led to the second, it would be quite wrong to say that a condition had simply been 'built into the primitive conjecture.'

Lakatos's method comes out looking better if we follow his ideas from the main text of *Proofs and Refutations*. There we find the following heuristic rule:

Rule 4. If you have a counterexample which is local but not global, try to improve your proof-analysis by replacing the refuted lemma by an un-falsified one. (Lakatos 1976: 58)

One of the students, Omega, then goes on to point out that there are two ways this rule may be implemented. In Poincaré's case, the first of these ways corresponds to restricting the validity of the theorem to manifolds for which all collections of submanifolds have the right intersection properties, thereby rendering it empty. The second way is to invent 'a completely different, more embracing, *deeper*, proof' (*ibid.*: 59). This is sound advice to be sure, but hardly helpful. One would expect a 'logic of discovery' to offer us a little more by way of insight as to how to find such a proof.

Furthermore, and this is this essential point, to the best of my knowledge, no other counter-examples did emerge to the large amount of theory contained in the 300 pages of the *Analysis Situs* papers. This is not to say that later mathematicians considered all his proofs complete, but simply that they proceeded not by discovering counter-examples to the proofs but by reformulating them. As regards the second stage of discovery of a theory, then, it would appear to be wrong to maintain that the Method of Proofs and Refutations describes the main dynamic of this period. Any attempt to 'rationally reconstruct' it as such would be doing unacceptable violence to history.

7.4 THE RESEARCH PROGRAMME STAGE

I shall now move on to discuss Lakatos's treatment of the transition to the third stage of discovery and the third stage itself. Most readings of Lakatos have focused on his thesis of the fallibilism of mathematical theories, that we can never know but only guess, never know that we have extended our knowledge only guess that we have. I hope to show here that this is an inaccurate appraisal of his ideas and that in many respects he inclined further towards the logicist pole than has been thought, in the sense that he believed proofs in axiomatic theories to be readily formalisable.

Steiner attributes to Lakatos the view that 'knowledge (meaning certainty) is impossible even in mathematics' (1983: 505), backing this up with a quotation from Lakatos which begins 'We never *know*: we only guess'. As the larger part of his disagreement with Lakatos seems to rest on this identification of knowledge with certainty, it would surely have been wiser to have been a little more careful in relying solely on this single quotation. When *knowledge* is used in its more usual sense it is clear that Lakatos does hold that it is something we can possess. Indeed, it is the growth of knowledge on which he bases his whole philosophy. Thus.

> It will take more than the paradoxes and Gödel's results to prompt philosophers to take the empirical aspects of mathematics seriously, and to elaborate a philosophy of critical fallibilism, which takes inspiration not from the so-called foundations but from the growth of mathematical knowledge. (Lakatos 1978b: 42)

Having equated knowledge and certainty for Lakatos, Steiner makes the claim that Poincaré's proof

> characterizes polyhedra in highly abstract terms to which the concepts of algebra apply. Lakatos has no counterexample here to offer (in fact, none have turned up); instead he complains that the proof is too abstract to have captured the original subject matter! (Steiner 1983: 504)

He then supports this claim by quoting Lakatos. The first thing to note about this quotation is that it is not taken from *Proofs and Refutations*. The second is that it does not contain such a complaint. Before proceeding to the quotation let us see whether the text of *Proofs and Refutations* gave Steiner any grounds for making such a claim. We find that Steiner might have been thinking of the comment made by one of the pupils:

Lambda: . . . This is no longer a theorem about polyhedra but about a certain set of multidimensional vector spaces. (Lakatos 1976: 118)

However, Lambda's later comments on the proof show that he has appreciated it as an improvement (*ibid*.: 120). On the other hand, it seems a fair bet to assume that the Teacher's comments are closest to Lakatos's own views and we find him saying during Epsilon's presentation of the proof:

Teacher: I like this reformulation which really showed the nature of your simple tools – just as you promised. You will no doubt prove Euler's theorem by the simple methods of vector algebra. Let us see your proof. (Lakatos 1976: 116)

and then later in reply to a criticism of the proof on the grounds that it is not final:

Teacher: But Epsilon never promised finality, only more depth than we had achieved earlier. He has now fulfilled his promise to produce a proof which explains both the Eulerian character of ordinary polyhedra and the Eulerian character of star-polyhedra at one blow. (Lakatos 1976: 120)

Later in the dialogue we find another pupil, Alpha, expressing worries that by translating vague terms into the language of a dominant theory

some essential aspects of the original vague concept may get lost. The new clear concept may not serve for the solution of the problem for which the old concept was supposed to serve. (Lakatos 1976: 122)

That Lakatos held the opinion voiced by Alpha may be seen from the quotation that Steiner *does* use to support his claim. This, however, merely states the widely held view that 'it is certain that we won't have any counterexample formalizable in the system assuming the system is consistent; but we have no guarantee at all that our formal system contains the full empirical or quasi-empirical stuff in which we are really interested and with which we dealt in the informal theory. There is no formal criterion as to the correctness of formalization' (Lakatos 1978b: 66). Lakatos then proceeds in the same paper to give some examples in which formalisations have been insufficient to deal with certain intuitive ideas pertaining to that field. That this assertion is unobjectionable to Steiner is shown by the fact that Steiner himself comes up with another example of this situation. But what then was the point he was trying to make? Lakatos did not say that all formalisations were doomed never to capture the content of their intuitive ancestors.[7]

---

[7] See (Lakatos 1978b: 67): 'Does all this mean that proof in a formalized theory does not add anything to the certainty of the theorem involved? Not at all . . . if we manage to *formalize* a proof of our theorem within a formal system, we know that there will never be a counterexample to it which could

Nor does he say that casting proofs in theories further removed from the intuitive subject matter cannot add to our knowledge of the result. This latter claim is the main purpose of Steiner's presentation of the modern version of the proof of Euler's formula at the conclusion of which he feels he has proved his point against Lakatos that 'contemporary treatments are superior to Euler's and Cauchy's in that they are more explanatory and explain more' (Steiner 1983: 521).

Had Steiner paid closer attention to Lakatos's writings he may have found himself disagreeing with Lakatos rather less strongly than he did and, seeing that he too has an interest in mathematical practice, may even have found useful ideas there. In defence of Steiner, it does seem to be especially easy to misread Lakatos's work owing to the difficulties, described by Ian Hacking (1979) in his review of Lakatos's *Philosophical Papers*, of understanding his underlying philosophical motivation.[8]

7.5 LAKATOS ON RIGOUR

A further common accusation made against Lakatos is that he has wrongly characterised standards of rigour as ever-changing. A typical example of this is the following remark made by the mathematician Saunders Mac Lane:

[T]here was (and, despite Lakatos, still is) a precise definition of "proof". (Mac Lane 1988: 325)

Now before countering this remark it is worth seeing what Mac Lane means by the notion of a rigorous proof:

For the concept of rigor we make a historical claim: That rigor is absolute and here to stay. The future may see additional axioms for sets or alternatives to set theory or perhaps new more efficient ways of recording (or discovering) proofs, but the

---

be formalized within the system as long as the system is consistent . . . if formalization . . . conforms with some informal requirements, such as enough intuitive counter-examples being formalized in it and so on, we gain quite a lot in the value of proofs.'

[8] Is this difficulty sufficient to account for the fact that we find such disagreements between reviewers of his work as 'Lakatos . . . espouses a correspondence theory of truth' (Steiner 1983: 508), and 'Lakatos's problem is to provide a theory of objectivity without a representational theory of truth' (Hacking 1979: 384)? Hacking does, however, go on to note (*ibid*.: 385n.) a doubt expressed by Feferman about his claim that Lakatos is not primarily concerned with characterising knowledge by how well it represents reality. Perhaps, matters could be resolved by thinking in terms of Hacking's notion of inherent-structurism and its extension to mathematics as I discussed in section 7.1. Then we can make sense of Lakatos's assertion – 'I think that the bulk of logic and mathematics is God's doing and not human convention . . . But in consequence I am a fallibilist not only in science, but in mathematics and logic as well' (Lakatos 1978b: 127) – as an avowal of anti-nominalism. Hacking tells me (personal communication) that he came to the idea of inherent-structurism during his time as Lakatos's assistant.

notion of a rigorous proof as a series of formal steps in accordance with prescribed rules of inference will remain. (Mac Lane 1986: 378).

If, for example, category theory were to become the standard language for mathematics, this should not be counted as a change in the standards of rigour – category theoretic proofs also proceed only via 'a series of formal steps'. Rather, the adoption of a new language will bear solely on the question of pragmatic efficacy.

That Lakatos did not dissent from Mac Lane's opinion we can see from the following passage:

> Up to now no informal mathematical theory could escape being axiomatized. We mentioned that when a theory has been axiomatized, then any competent logician can formalize it. But this means that proofs in axiomatized theories can be submitted to a peremptory verification procedure, and this can be done in a foolproof, mechanical way. (Lakatos 1978b: 66)

Indeed, Lakatos appears to be saying something stronger than Mac Lane, who realises that the complete formalisation of a proof is an extremely difficult and laborious process.

Having deflected this criticism, I can now explain the cause of my disagreement with Lakatos. Later in the same article he warns again of what can happen if a pre-formal theory is axiomatised too early. There follows a series of what-if scenarios and the information that 'even after a theory has been fruitfully axiomatized, there may arise issues which can bring about a change in axiomatization' (*ibid*.: 68). So what is the worry on his part? Life and uncertainty seem to go on even after axiomatisation. But here we come to the crucial point:

> While in an informal theory there are unlimited possibilities for introducing more and more terms, more and more hitherto hidden axioms, more and more hitherto hidden rules in the form of new so-called 'obvious' insights, in a formalized theory imagination is tied down to a poor recursive set of axioms and some scanty rules. (1978b: 68)

The impression given here and in other places in his writings is that the excitement of bold speculations and illuminating intuitive insights disappears on reaching the relative safety of axiomatisation. Thus:

> [i]t is important to realize that most mathematical conjectures appear before they are proved; and they are usually proved *before* the axiomatic system is articulated in which the proof can be performed in a formalized way. (Lakatos 1978b: 96)

While the first claim is surely trivial, one has only to leaf through a contemporary book such as *Open Problems in Topology* (van Mill and Reed 1990) to realise that the second claim is completely false. In this book we find over 1,000 unsolved problems of the different branches of topology, many in the form of conjectures and the vast majority only expressible after axiomatisation had taken place, gathered together to help prospective researchers choose from a list of unanswered problems deemed important by the community.

The implication that the creativity and uncertainty are largely confined to the pre-axiomatic stage is reinforced by such comments as '[a]nalysis was not any more a venture into the unknown; it was an exercise in mobilizing and ingeniously connecting the various parts of the known' (Lakatos 1978b: 100).

I certainly would not wish to disagree with Lakatos that each theory needs intuitive motivation; there are too many cases of theories that have been developed simply by dropping an axiom of an established theory and which, owing to lack of sufficient motivation, have not led to anything worthwhile. But by asserting that '[t]here is indeed no respectable formal theory which does not have in some way or another a respectable informal ancestor' (*ibid.*: 62), he appears to be saying more than this. He seems to be implying that if you take a body of theory presented in terms of some modern axiomatic theory and trace back its ancestry, you will arrive at a period when the concepts of the later system could be seen condensing out of the undefined and nebulous proto-theory. I challenge anyone who holds this view to try to find an informal ancestor of an Eilenberg–Mac Lane space or a spectral sequence dating from the pre-axiomatic stage of algebraic topology.

The polyhedron concept is one that lies close to basic human experience: we play with dice, we receive postcards of the Egyptian Pyramids, and we eat bars of Toblerone. However, the supply of such low-level concepts is surely limited, as those who attempt popularisations of mathematics well know. Explaining modern physics to the man in the street is somewhat simpler in that he can easily imagine two particles colliding and producing new particles, or can strain his geometric imagination to think of space-time as a four-dimensional manifold. But this paucity of intuitions among the mathematically uneducated does not mean that trained research mathematicians works only with the formalism of a theory. There is a common misconception present in those who have reached the end of *their* imagination that everyone else must be in the same boat, merely relying on the formalism to indulge in some pointless game. As Thurston, the geometer, puts it: 'One person's clear mental image is another person's intimidation' (Thurston 1994: 164).

The idea that mathematicians have an intuitive feeling of the behaviour of objects they try to define is surely right, but the process of discovery today involves as it ever did the struggle to find a good or the 'right' definition. Even after this has occurred the dialectical process continues as novel proof techniques emerge and are then better understood. Peter Johnstone, in the introduction to his book *Topos Theory*, is apologetic for having contributed to the 'concreting-over' of the foundations of the basic theory of elementary toposes, seeing this as a bad thing because 'it is vital to the health of a subject as basic as topos theory that its fundamental tenets should be the subject of continual review and improvement' (Johnstone 1977: xvi), yet he does, of course, mean this review to be carried out in a rigorous fashion. This improvement of 'fundamental tenets' is not meant to imply the overthrow of the axioms defining an elementary topos, which have not changed from the first.[9] Rather, it refers to the way the theory is organised, the relative importance given to different subclasses of toposes and the morphisms between them, the preference for certain proof techniques, etc. The point here is that even after the arrival of rigour there is still room for debate; concept-stretching continues to occur.

Lakatos missed one of the essential roles of axiomatisation as Kreisel and MacIntyre (1982: 233) intimate: 'The use of axiomatic analysis as a proof strategy does not seem to be well known to people writing on heuristics, like Pólya, nor to those in the education business.' They see its significance less for the greater generality of the theorem proved or any idea of greater *precision*, but more as 'a strategy both for finding and remembering proofs' (*ibid.*: 232). A similar view is voiced by Eilenberg and Steenrod (1952: x–xi):

The great gain of an axiomatic treatment lies in the simplification obtained in proofs of theorems. Proofs based directly on the axioms are usually simple and conceptual . . . Successful axiomatizations in the past have led invariably to new techniques of proof. The present system is no exception.

Axiomatisation has not hindered but aided the creativeness of the best modern mathematics by helping to disentangle theories from the contingent circumstances in which they were discovered:

It is as if you took a man out of a milieu in which he had lived not because it fitted him but from ingrained habits and prejudices, and then allowed him, after thus setting him free, to form associations in better accordance with his true inner nature. (Weyl 1951: 465)

---

[9] In point of fact, one of Lawvere and Tierney's axioms was later found to be redundant.

Thus axiomatisation facilitates interaction between apparently different areas of mathematics, an essential feature of the twentieth century, and has helped to overcome the compartmentalisation of earlier centuries.

We can gain a clearer understanding of what is involved here by approaching these issues from a different angle, namely, by studying a criticism expressed by Davis and Hersh about the inclusion of comments made by the editors of *Proofs and Refutations*, John Worrall and Elie Zahar, in a series of four footnotes they added to the text. Davis and Hersh (1981: 354) complain that Worrall and Zahar have misunderstood Lakatos's message that it is an error to identify 'mathematics itself (what real mathematicians really do in real life) with its model or representation in metamathematics, or, if you prefer, first-order logic'. Worrall and Zahar had maintained that 'first order logic has arrived at a characterisation of the validity of an inference which (relative to a characterisation of the "logical" terms of a language) does make valid inference essentially infallible' and that Lakatos, having come in later years 'to have the highest regard for formal deductive logic' would have given up on his claim that we should 'give up the idea that our deductive, inferential intuition is infallible' (Lakatos 1976: 138).

This discrepancy in the evaluation of Lakatos's ideas can be explained by Davis and Hersh imagining Lakatos's understanding of *modern* mathematics to be in agreement with theirs, that is, that his discussion of the growth of informal concepts was intended to extend to the twentieth century and up to the present day. In that Lakatos was the only philosopher of mathematics voicing the kind of opinion they wanted to hear, it is understandable that they imagined his views to be closer to theirs than in fact they were, and so thought that Worrall and Zahar's additions ran squarely against Lakatos's thinking. It appears, however, on closer inspection of Lakatos's writings that it is Worrall and Zahar who are largely correct in their interpretation of him.

Now Davis and Hersh know, being practising mathematicians, that theirs is a difficult profession, one which is full of uncertainty, but in this article they restrict the range of this uncertainty to the realm of proofs, a typical example of which appearing in a mathematical journal of today is certainly not written in a completely formal style. Yet, whereas we have seen earlier Lakatos say of proofs in modern theories that 'any competent logician can formalize [them]' and thus that they 'can be submitted to a peremptory verification procedure', Davis and Hersh maintain that '[a] real proof is not checkable by a machine' (1981: 354) and believe that Lakatos is with them on this issue:

Lakatos applied his epistemological analysis, not to formalized mathematics, but to *informal* mathematics in process of growth and discovery, which is of course mathematics as it is known to mathematicians and students of mathematics. (1981: 347)

However, Lakatos reckons that there are two senses of the word 'informal' as applied to proofs:

But what about an informal proof? Recently there have been some attempts by logicians to analyse features of proofs in informal theories. Thus a well known modern text-book of logic says that an 'informal proof' is a formal proof which suppresses mention of the logical rules of inference and logical axioms, and indicates only every use of the specific postulates.

Now this so-called 'informal proof' is nothing other than a proof in an axiomatized mathematical theory which has already taken the shape of a hypothetico-deductive system, but which leaves its underlying logic unspecified. At the present stage of development in mathematical logic a competent logician can grasp in a very short time what the necessary underlying logic of a theory is, and can formalize any such proof without too much brain-racking.

But to call this sort of proof an informal proof is a misnomer and a misleading one. It may perhaps be called a quasi-formal proof or a 'formal proof with gaps' but to suggest that an informal proof is just an incomplete formal proof seems to me to be to make the same mistake as early educationalists did, when, assuming that a child was merely a miniature grown-up, they neglected the direct study of child-behaviour in favour of theorizing based on simple analogy with adult behaviour. (Lakatos 1978b: 62–3)

Thus, while Davis and Hersh imagine that Lakatos's account of informal mathematics extends to present-day proofs in established branches of mathematics, surely the majority of the estimated 200,000 produced each year, Lakatos himself wishes to count them as 'almost formal' or 'quasi-formal'.[10]

If we imagine a spectrum of opinion about the reasons for the acceptance of proofs in *modern* theories as being correct ranging from an extreme left which holds that this is decided by some social decision process of the mathematical community, to the extreme right which would say that a proof is acceptable only if it is possible to formalise it in a truth transmitting calculus, then we find that while Worrall and Zahar want to have Lakatos seen as further to the right, at least in his later years, the majority of commentators, and many of these disapprovingly, interpret him as sitting towards the left. It is no doubt impossible to locate Lakatos precisely on this spectrum, but I hope I have shown that Worrall and Zahar's interpretation

---

[10] We could attempt to make a distinction between the use of the terms 'axiomatisation' and 'formalisation', but need not in this discussion as Lakatos claims to use the terms interchangeably (see *ibid*., p. 67).

is not unfaithful. What is novel in Lakatos is that, unlike most people who see formalisation as unproblematic, he does not take it as an end in itself and worries about the possibility of its excluding fruitful ideas.

Further testimony to the correctness of this line is offered by the way that Lakatos tends to favour the syntactic over the semantic aspects of theories. If we read his characterisation of the 'hard core' of the research programme of algebraic topology it is given in terms of its axioms (1978b: 96).[11] Now, ignoring for the moment the frequent assertions of mathematicians that the boundary between two branches is ill defined, we can say that modern algebraic topology may be partially characterised by its aim of resolving problems in topology by assigning algebraic objects to topological objects in an invariant (up to homeomorphism) way and thus reducing the original problems to easier algebraic ones. However, it is far from clear how such a characterisation could be described solely in terms of axioms. The axioms used by algebraic topologists include those defining types of topological spaces, those defining types of algebraic objects, such as Abelian groups and graded rings, and those defining processes going from one type of object to the other. The Eilenberg–Steenrod axioms play the role of definitions picking out what their authors decided was an important class of collections of maps (functors) indexed by the integers from a suitable category of pairs of topological spaces to the category of Abelian groups. The principal reason for doing this was to facilitate the proof of general results by focusing on the essential characteristics of the, until then, motley collection of homology theories. Many of the axioms were properties of these homology theories which had been proved for them individually.

As I mentioned earlier, homology theory was far from being the whole of algebraic topology at that time, but even if we restrict ourselves to this subbranch, we find that the appearance of new concepts did not cease after 1952. Shortly after this date new collections of functors were discovered satisfying all the Eilenberg–Steenrod axioms except one – the dimension axiom, which requires that the homology groups of the space consisting of

---

[11] Other commentators have noted this aspect of Lakatos's thought. Eduard Glas (1989: 168) describes Felix Klein's programme as 'the best example of progress through generalization and consolidation' likening it to Lakatos's methodology, but he then adds that 'proofs and refutations typically belong to propositions, and Klein was not primarily concerned with the proofs of propositions but with the development of models'. Kitcher's view of explanatory progress leads him to recognise '*the need to break away from concentration on accepted statements* (a feature of logical empiricism that survives in Lakatos and Laudan) *and to focus on the ways in which statements are* used *in answering questions*' (Kitcher 1993: 112n., author's emphasis). There was always the fear in Lakatos's mind that too many concessions on this issue would lead to the introduction of something like Polanyi's 'know-how', which Lakatos felt to be irrelevant to his programme of explaining the development of theories in the 'Third World'.

a single point be zero in all dimensions except dimension zero. This axiom was introduced, the authors claim, to 'insure that the dimensional index shall have a geometric meaning' (Eilenberg and Steenrod 1952: 12). These functors, which include K-theory, now known as *generalised* homology theories, are seen as very important developments and, despite their not satisfying the dimension axiom, still have a strong geometric meaning. Should we say that this premature axiomatisation by Eilenberg and Steenrod might have delayed, or even prevented, the discovery of these new homology theories? On the contrary, it is impossible to imagine the dramatic progress which occurred in algebraic topology through the 1950s without the spread of this new conceptual approach promulgated by means of the publication of their book.

Returning to Davis and Hersh, we can now see that they were missing the most essential point. They know that formalising a proof contributes nothing to its comprehensibility, that mathematical understanding is not transmitted without sufficient motivation, and that the logical form of a proof fails to capture many of its important features,[12] yet in this paper they attend solely to the notion of the reliability of proofs. Given the stabilisation that has occurred in the idea of what constitutes a rigorous proof, this gives them little room for manoeuvre against their formalist adversaries. Elsewhere we find that they, or at any rate Hersh, recognise that uncertainty is more commonly to be found in aspects of modern mathematics other than the validity of proofs:

there is an amazingly high concensus in mathematics as to what is "correct" or "accepted". But besides this, and equally important, is the issue of what is "interesting" or "important" or "deep" or "elegant". These aesthetic or artistic criteria vary widely, from person to person, specialty to specialty, decade to decade. They are perhaps no more objective than aesthetic judgments in art or music. (Hersh 1991: 131–2)

These criteria are precisely what we shall be studying in chapters 8 and 9. Mathematicians today, aware of the volume of production of their colleagues, are far more concerned that their work will be ignored through lack of interest shown to it than through any fear that it will be found incorrect. Mac Lane (1986: 441) has a list of values for a piece of mathematics similar to Hersh's in which he too avoids the use of the word *true* in favour of *correct*, and then adds the epithets: responsive, illuminating,

---

[12] This is one of the interesting points made by Kreisel and MacIntyre (1982). They recommend (1982: 236) the perusal of essay reviews or lectures to learned societies by mathematicians as an antidote to misconceptions on this score.

promising, and relevant. However, he differs from Hersh, rightly in my view, in the belief that there is something more objective to say about these qualities.

7.6 THE JAFFE–QUINN DEBATE

A further twist in this story about the place of rigour comes in an article by Arthur Jaffe and Frank Quinn (1993). These two mathematical physicists are worried about the emergence of a new hybrid discipline lying somewhere between mathematics and physics. As we have seen in chapter 5, theoretical physicists often find themselves today dealing with theories whose consequences are beyond the range of experiment and so they are turning to mathematicians to fill the role of experimentalist in that their speculations may lead them to, say, a description of representations of the 'monster' sporadic group using vertex operators in Kac–Moody algebras. While the authors are wholeheartedly in favour of mathematicians drawing inspiration from ideas in physics, they are concerned that much of the mathematical work emerging has not been established in the traditional rigorous fashion with the following consequences to the practice of mathematics:

(1) Theoretical work, if taken too far, goes astray because it lacks the feedback and corrections provided by rigorous proof.
(2) Further work is discouraged and confused by uncertainty about which parts are reliable.
(3) A dead area is often created when full credit is claimed by vigorous theorizers: there is little incentive for cleaning up the debris that blocks further progress.
(4) Students and young researchers are misled. (1993: 8)

Here we have, if you like, the reverse of Lakatos, in the sense that Jaffe and Quinn are asserting that *delay* in the introduction of rigorous definitions and proofs to a theory, if not properly signalled, may well have a pernicious effect. Some of the best speculative ideas are not fulfilling their potential, they say, because their discoverers are eschewing rigour. In other words, an important part of the dialectical process is being missed in that good intuitive ideas, which are often the material for the most fruitful variety of rigorous exploration, are being drowned in a sea of conjectures from which they may be extracted only by great effort.

Jaffe and Quinn illustrate their thesis with examples of lack of rigour slowing progress and, in particular, in response to Dieudonné's query as to why it took so long after Poincaré's work for algebraic topology to establish itself as a subject (*ibid.*: 36n.), allege that the result of Poincaré's 'reckless' method was 'a dead area which had to be sorted out before it could take off' (*ibid.*: 7). While there is no doubt a modicum of truth in this claim, we ought I feel to offer a defence on Poincaré's behalf and through this provide a vindication of Lakatos's interest in the less-than-formal, intuitive side to mathematical reasoning.

A first point to be made is that in that era, fifteen or so years was not a long period of time for ideas in a new area to be taken up. After all, the interval from Betti to Poincaré was even longer. Many branches of geometry were especially slow to crystallise. By 1900, after a century of the most violent overthrows of geometrical understanding, the definition of a differentiable manifold and of a topological space had not yet stabilised, and would not do so for another decade. Thus:

[i]t is quite difficult for us to understand the point of view of the mathematicians who undertook to tackle topological problems in the second half of the nineteenth century: When dealing with curves, surfaces, and, later, manifolds of arbitrary dimension, with their intersections or their existence when submitted to various conditions, etc., they relied exclusively on "intuition", and thus followed – with a vengeance – in the footsteps of Riemann, behaving exactly as the analysts of the eighteenth or early nineteenth century in dealing with questions of convergence or continuity! (Dieudonné 1989: 15)

Rigour in algebraic topology, according to Dieudonné, comes only with Brouwer's work in around 1910, but the question of which objects form its most appropriate domain has continued to be asked right up until recent times.[13]

Now we are used to the idea of infinite dimensional spaces it may be hard for us to remember the novelty of working in dimensions higher than 3. In the introduction to *Analysis Situs* Poincaré went to the lengths of justifying the need for such a theory by remarking that while Riemann had classified algebraic curves according to their genus by classifying closed real surfaces using topological methods in two dimensions, the classification of algebraic surfaces and their birational transformations was going to require topological considerations in five dimensions.

We should notice in addition the originality and depth of the ideas emerging from the *Analysis Situs* papers. It is generally recognised that a

---

[13] It seems fair to say that the geometric branches of mathematics that have taken longest to 'settle down' are the ones most likely to bring about reconceptualisations of mathematics. It was algebraic topology which led to the formulation of the first ideas in category theory and it was in algebraic geometry, for which Dieudonné dates the advent of rigour as late as 1950, that the notion of a topos first emerged.

majority of the concepts employed in algebraic topology until 1925 and many of those appearing after this date have their seed in Poincaré's work. It certainly took a rare combination of talents to produce these papers and, if a talent for rigorous work was not among these, then this is surely quite understandable.

Finally, it is interesting to note some of the mathematicians who *did* follow Poincaré. The four most important of these before 1925 were, perhaps, Brouwer, self-taught and isolated in Holland, who did have a talent for rigour but restricted himself to a small fragment of what his new concept of simplicial approximation allowed him to explore, and the three Princeton topologists, Alexander, Veblen and Lefschetz who, as residents of a country with a shorter history of mathematical research, were not so constrained by tradition. In this respect, it is interesting to note that in France where analysis continued to reign supreme, nobody took up algebraic topology until Jean Leray in the 1940s. We might have imagined that Poincaré's students would have carried on their teacher's work and might well find that a historical account of Poincaré's role as a supervisor would yield a more complete answer to Dieudonné's question. Indeed, we find Dieudonné himself (1975a: 52) making a comparison between the working styles of Poincaré and Gauss when he notes that both of these mathematicians 'had very few students and liked to work alone'.

Perhaps we can arrive at a synthesis of Lakatos and Jaffe–Quinn's positions. Mathematical research is to a great extent a search for ideas. There are times when a domain needs an injection of rigour and times when it can benefit from some speculative thinking. From the opposite point of view, either activity may block the production and development of ideas. As with comedy, it's all in the timing.

7.7 CONCLUSION

I would like to end this chapter with a tentative suggestion as to why Lakatos had such a poor opinion of axiomatisation. Herbert Breger (1992) has argued that the establishment of the modern axiomatic paradigm by Hilbert, Zermelo and others occurred via the overthrow of the earlier paradigm of *extreme Platonism*. Despite the fact that it was possible to be a member of this new paradigm and still be a kind of Platonist and despite Hilbert's profession that he was preserving the old order, according to Breger, a central belief of the older paradigm was given up. Whereas for Hilbert, consistency of axioms for a system was sufficient reason to say that that system existed, for the extreme Platonists 'a definition does not constitute an object, it just points at an already pre-existing object.' (*ibid*.: 256). Finsler was an adherent of the old paradigm still working in the late 1920s on an axiomatisation of set theory in the Cantorian tradition. However, this style of axiomatisation was very different from that of the Hilbertians. His is maximal in the sense that he is describing the totality of all sets as that system which satisfies certain axioms and which cannot be extended; theirs is minimal in the sense that there are no sets which cannot be derived from the axioms. If we are right about Lakatos's inherent-structurism, it would be consistent for Lakatos to see axiomatisation in the Hilbertian sense, which is of course the one that has been largely adopted by the mathematical community, as somehow restricting the freedom of mathematicians to capture all the mathematical reality there is.

If this sketch is roughly along the right lines, it would account for the refreshing novelty of Lakatos's philosophy of mathematics. Despite my criticisms of him, Lakatos has played a vital role in redirecting our attention to the evolution of mathematical theories, in particular through his treatment of concept-stretching. However, I have argued that he misunderstood the range of applicability of his ideas. He has claimed too much for them in the sense that the method of proofs and refutations has at no time been the most important mechanism for progress in mathematics, while he has claimed too little for them by restricting himself to pre-axiomatic theories and in doing so made too much of the division between axiomatised theories and their informal ancestors. Mathematical concepts may still be generated today via proof attempts and uncertainty still thrives in the guise of such problems as the choice of direction one's research should take or the 'right' way a theory should be generalised.

Insofar as the method of proofs and refutations focuses too closely on the level of proofs, it demonstrates its limitations. What we need is an account of explanatory and conceptual progress which goes beyond the bounds of the collection of accepted statements of a theory. Let us now see whether a methodology of mathematical research programmes might fare better.

CHAPTER 8

Beyond the methodology of mathematical research programmes

8.1 INTRODUCTION

It is well known that, prior to his untimely death, Lakatos had hoped to return to the philosophy of mathematics bringing with him there some of the constructions he had formulated during his study of the development of the natural sciences. The problem which had faced him in the philosophy of science was to rectify Popper's depiction of scientific methodology taking account of Kuhn's criticisms, yet avoiding what he saw as the relativist pitfalls of the Kuhnian approach. He inherited from Popper an appreciation of the heuristic role of metaphysics in scientific theorising, which had marked Popper's difference from the Logical Positivists, and indeed placed a greater emphasis on this role by allowing room for it within the structure of a research programme.[1]

We may observe, then, that as he switched from mathematics to science, the targets of his philosophical criticism changed from logicism and formalism to positivism and Kuhnian relativism. This should lead us to wonder which threats he would have been warding off with a methodology of *mathematical* research programmes. Two possibilities suggest themselves. First, he might have wanted to use this methodology to intensify his attack on logicism and formalism begun in *Proofs and Refutations* and continued in other works. Second, seeing that in the 1970s there was no apparent relativist threat as regards mathematical knowledge, he might also have wanted to create a prophylactic against the future appearance of such a threat.

Let us take these targets in turn. Just as he constructed his methodology of scientific research programmes to oppose the Positivists' notion that the philosophically significant part of the scientific enterprise was restricted to a collection of scientific statements logically analysable into observation

---

[1] 'Whereas Popper acknowledged the influence of metaphysics upon science, I see metaphysics as an integral part of science' (Lakatos 1978a: 148n.). As we shall see, however, Lakatos was not so successful in distancing himself from other Positivist principles.

statements, so he might have used mathematical research programmes to oppose the logicist and formalist positions by underscoring the heuristic and value-laden components within them. However, to be in a position to see such programmes as dynamic progressive entities he would have had to adopt a far more positive opinion of axiomatisation. Lakatos made the bold proposal that something might be missed by the formalisation of informal thought, yet he needed to understand how axiomatisation might act to support the generation of a more refined informal thought, rather than, as we saw in chapter 7, as marking the termination of the creative process of research. Had he realised this, he would have been able to extend greatly his fallibilist conception of axiomatised mathematics beyond the limited sphere of logical falsification (when contradictions appear) and heuristic falsification (when aspects of the prior informal theory had not been captured).

As for the second of Lakatos's possible targets, the production of a methodology of mathematical research programmes would have a bearing on ongoing debates within what is sometimes termed the post-positivist philosophy of mathematics. Here, a central bone of contention is the issue of what mathematicians have termed the *unity* (Michael Atiyah) or *connectivity* (Saunders Mac Lane) of mathematics. Is modern mathematics a loose association of disparate theories related to, as Mehrtens puts it, 'heterogeneous specific problems' (1990: 20) with a 'form of communication . . . [which] tends to sharpen internal boundaries between specialties' (quoted in McLarty 2000: 271)? Or, should we see these claims of Mehrtens as arising from an 'excessively direct transcription of common views of postmodernism into the history of mathematics' (McLarty 2000: 278) and instead strive to account for the profound interconnectedness of modern analysis, geometry, topology and number theory and their current use in mathematical physics and elsewhere? There are indications that Lakatos feared that the former was the more accurate picture with the concomitant worry that much of modern mathematics was degenerate. Presumably, however, he would have expected a methodology of mathematical research programmes either to reveal the rationality of modern mathematical practice or else to point to ways of enhancing it.

A handful of attempts have been made to construct a methodology of mathematical research programmes. Hallett (1979), for example, finds no great obstacles to this task. More recently, we find Glas agreeing that science and mathematics share a similar Lakatosian methodology, yet he feels that this needs to be enriched by sociological considerations:

This adapted version of the methodology has been shown to be adequate to account retrospectively for the development of theories that depend on a shared hard core of assumptions, but nothing more. In particular, it does not account for the genesis of a new programme nor for the transition from one programme to another. (Glas 1995: 241–2)

So Glas requires a 'sociohistorical perspective in the Kuhnian vein' (*ibid*.: 242) to complement Lakatos's methodological perspective, and this he outlines in the same paper.

I am far more sympathetic to sociologically inspired accounts of the development of mathematics than Lakatos would have been, and I believe much important work has been done in this area yet, like McLarty, I feel there is a tendency to exaggerate the fragmentation of modern mathematics. Lakatos surely failed to convince us that he had discovered the laws of growth of knowledge in some autonomous 'Third World', but testing the aptness of research programmes in the history of modern mathematics may enable us both to decide on the degree of connectivity and to weigh up judiciously the factors determining what the mathematicians of any era deem to be salient.

Glas's 'adapted version of the methodology' maintains the Lakatosian picture of well delineated non-interacting research programmes and this lends support to a vision of the course of mathematics being governed to a very appreciable extent by external interests, such as the promotion of mathematicians' careers.[2] A concern here is that if the depiction of the development of research programmes does not include consideration of the interaction between programmes, we shall overlook the fact that large tracts of mathematics may be described in terms of a small number of powerful principles (see, for example, the range of harmonic analysis as presented by George Mackey in his (1992)).

On a speculative note, the existence of such principles offers us, I believe, a clue as to the relationship between our cognitive capacities and the comprehensible features of the world. This brings me to a further point that I wish to make about Glas's presentation. I would suggest that even his socio-historical addition, while important, will not suffice to account

[2] Discussing Monge's Central School of Public Works, Glas claims that '[t]he thorough educational reforms effected in this prestigious institution (which later would still inspire Klein) produced a generation of students who stuck to his "paradigmatic" teachings because they embodied the special competences and proficiencies on which their further careers depended' (Glas 1995: 240). With the vast number of mathematics departments and journals in existence today, and the possibility of rapid communication and travel, we might wonder how relevant to the present-day situation are the factors Glas has isolated.

for 'the genesis of a new programme'. Clearly, for a piece of mathematical improvisation to become accepted it cannot deviate too far from the internalised norms and values of a sufficiently powerful section of the mathematical community, yet constraints on what may be *understood* must arise not just from our condition as culturally constituted beings, but also from our embodied condition. Better still, these conditions should be considered as interrelated.

What must also not be neglected is what we might call the *ludic* quality of mathematical research. Mathematicians after immersing themselves in various theories will devise problems and 'play around' with novel techniques and concepts in their attempts at solving them, based on their desire to increase their understanding. To some degree, and I deem it relatively large (especially in the twentieth century), important ideas have arisen through mathematicians being allowed to take up the challenge of clarifying and extending theories with little need of extra-mathematical justification. The question remains whether there exists some characterisation of the objective mathematical importance of a technique or concept. This we shall in address in this chapter and chapter 9.

In this chapter I shall attempt to determine the extent to which the research programme schema could be made to work for mathematics in the modern era. I shall begin by examining the problems of delineating research programmes in an age where the pace of mathematical research is so much faster than that obtaining before, say, 1850. This I shall attempt for algebraic topology, a minor part of whose rise into a research programme we have seen Lakatos describe in *Proofs and Refutations*. One hypothesis that will be thrown up from this section is that rivalry between research programmes concerns high-level issues. This I shall illustrate in the following section by considering a battle between Dedekind and Kronecker and their followers as to whose theory was to succeed Kummer's number theory.

The principal point that emerges from this study is that mathematics works at many levels, and that the one-level analysis provided by the methodology of research programmes is insufficient. This thesis will be examined in the context of a brief critique of some ideas of Giulio Giorello. Similar criticisms have been made by Teun Koetsier, who decided in his (1991) to turn to Larry Laudan's scientific research traditions for inspiration. Koetsier has realised that a more complex schema is needed to describe modern mathematics, but I shall proceed to show that he has not gone far enough. Such complexity – in particular the interweaving of theories – is not, I believe, due to the failure on my part to 'rationally reconstruct' away from the details of actual history. Instead, it is an integral feature of the dynamics of modern mathematics which accounts for the nature and rapidity of its development. Finally, I shall make some comparisons between my ideas and those of Penelope Maddy.

8.2 PUTTING THE METHODOLOGY OF MATHEMATICAL RESEARCH PROGRAMMES TO THE TEST

The components of a Lakatosian scientific research programme are: the hard core, the positive and negative heuristic, and the protective belt. According to Lakatos, the hard core of a research programme is a collection of laws that the adherents to the programme are willing to hold on to at any price. The reticence in allowing the hard core to be refuted is formulated as the negative heuristic. The positive heuristic is a plan to guide the development of the programme along lines which should eliminate some of the major anomalies that the programme faces and which may attend it from its birth. It 'consists of a partially articulated set of suggestions or hints on how to change, develop the 'refutable variants' of the research-programme, how to modify, sophisticate, the 'refutable' protective belt' (1978a: 50). This protective belt is a body of auxiliary assumptions and theories which can always be made to take the blame if an experimental finding appears to contradict a consequence of the theory.

Lakatos's prime example of a research programme is given by Newtonian mechanics, the hard core here comprising Newton's Laws of Motion and the Law of gravitation. According to the plan of the positive heuristic, Newton was to start by considering a planetary system with a fixed point-like sun and a single point-like planet. This assumption he recognised to be incompatible with one of the elements of his hard core, namely, the law stating that every action has an equal and opposite reaction, whose admission requires the complication of considering the system as revolving about the common centre of gravity. According to Lakatos's story, Newton then added other planets, but ignored interplanetary forces, introduced ball-like planets and sun, allowed the planets to spin, and so on. These changes were not motivated by experimental disconfirmation, but rather by foreseen theoretical contradictions already present within the programme.

Research programmes are defined by Lakatos as progressive in three different ways: (a) they are *theoretically* progressive if each modification of the protective belt 'leads to new unexpected predictions', (b) *empirically* progressive if some of these predictions are verified, and (c) *heuristically* progressive if the modifications are 'in the spirit of the heuristic'. In all these senses Newtonian mechanics constituted a progressive research programme over many decades.

One of Lakatos's students, Elie Zahar, introduced a modification to Lakatos's initial scheme. This was to alter the definition of empirical progress by saying that a research programme receives empirical support if the new theory leads to the understanding of novel facts, taking a fact to 'be considered novel with respect to a given hypothesis if it did not belong to the problem-situation which governed the construction of the hypothesis' (Zahar 1973: 1). In other words, the phenomenon which is explained need not be novel, so long as the theory was not devised with this phenomenon in mind. Thus, Copernicus's accounting for the retrograde motion of Mars is to count as empirical progress despite the fact that this phenomenon had been first observed centuries earlier.

I shall not comment at any length on the problems with the methodology of *scientific* research programmes (see, for instance, Chalmers 1982), but we should note the tension produced by considering science in terms of the activities of individual scientists and their communities and in terms of the evolution of the 'third world'. On the one hand, Lakatos wants to use the history of science and mathematics to tell a story of the progress of theories, the alienated products of human activity which inhabit Popper's 'World 3', separate from the worlds of physical reality and human consciousness. Thus, the methodologist needs to 'rationally reconstruct' historical episodes, abstracting away from the contingencies and accidental features of the discovery process, so as to describe the laws governing the growth of the occupants of this Third World. On the other hand, to determine empirical progress, in Zahar's modified sense, he must, as Feyerabend points out (1976: 222n.), make detailed investigations into the heuristic reasoning that lies behind the discovery of a theory. Only such detailed study into the creative thought processes of the scientists concerned, involving extensive investigation into their journals and correspondence with colleagues will allow us to establish whether the scientist was thinking of a particular phenomenon when he formulated a theory. This for Zahar is no drawback given his belief that 'the process of discovery is much more rational than it appears at first sight' (Zahar 1983: 244–5), that creativity is not the irrational, 'a-ha' process it has often been considered to be, and that many of the steps which lead to a theory's discovery are also employed in its justification. However, this would appear to take us beyond mere 'Third World' considerations.

In chapter 7, I described how Lakatos thought that the hidden lemmas which emerged and solidified during the process of proofs and refutations became the axioms which constituted the hard core of the newly established research programme. This idea, I argued, seems difficult to defend for algebraic topology given the speed with which homology theories (termed *extraordinary*) contradicting one of the axioms (the dimension axiom) were admitted after axiomatisation. Moreover, *cohomology operations* which relate the products of these axiomatised cohomology theories were discovered to be of use in settling more subtle topological problems.

The way I favour resolving this anomaly is to give up having axioms as the hard core. If we wish to find something that has remained truly invariant through the history of algebraic topology, it is the belief that it is possible to solve (all) important topological problems, many of which may be formulated as 'extension' or 'lifting' problems, by associating algebraic objects and mappings to topological spaces and their mappings in a 'natural' way, with the ultimate aim of being able to classify completely a suitably defined class of topological spaces along with the mappings between them.[3] To this may be added the belief that powerful connections may be made between the structure possessed by a space, i.e., the fact that it is an algebraic variety, differential manifold or whatever, and the invariants of the underlying topological space. A well-known example of such a connection relates the integral of a surface's curvature to its topological genus.

What this amounts to is a shift of perspective from seeing a mathematical theory as a collection of statements making truth claims, to seeing it as the clarification and elaboration of certain central ideas by providing definitions to isolate classes of relevant entities and ways of categorising and organising information about these entities. Further examples would describe harmonic analysis as based on the idea that functions can be expressed as sums of harmonics and algebraic geometry as the study of the solutions of polynomial equations. In each example there is a kind of creative vagueness to the central idea. For certain periods, the entities of a theory may change little while the apparatus used to investigate them varies; at other times, the range of the entities may be broadened.[4]

---

[3] As I mentioned in chapter 7, Eduard Glas had moved in this direction by characterising the research programme's hard core in terms of the recognition of the importance of the Euler formula. I have argued that this imputes to Lakatos more than he said, since I read him to take the constituents of the hard core to be axioms. This reading of mine is consistent with his taking the hard core of a research programme in physics to be composed of laws.

[4] As, for instance, when algebraic geometry allowed the coefficients of polynomial equations to lie in fields other than the real or complex numbers. My use of the word 'apparatus' points in the direction of exploring mathematics in terms of machines as in the interesting paper of Jean-Pierre Marquis (1997). One place I disagree with Marquis in this paper concerns the fixed distinction between instruments and natural kinds. In his examples from algebraic topology, algebra is represented as providing the instruments to probe 'natural' topological spaces. However, homological algebra can be turned on

Is this modification brought on us by some fundamental difference between mathematics and science? A Newtonian if asked whether he thought the law of gravitation and Newton's three Laws of Motion were true, might well have answered affirmatively, as would a seventeenth-century mathematician have responded to a similar question concerning the axioms of Euclidean geometry. But were we to ask a mathematician today whether he considered a system of axioms to be true, he would most probably tell us that he thought it had interesting models or that it described an important construction. Perhaps the distance between the modern mathematician and scientist in this respect is not so great in that the physicist is likely nowadays to say that she thinks her theory a good model of some aspect of the universe. Thus, Lakatos ought to have allowed a wider notion of hard core to allow research programmes any chance of success in accurately portraying science. In retrospect, the hard core of the Newtonian theory of planetary motion should also be seen to include beliefs which employ the notion of a *model*: massive particles are well represented by points in a three-dimensional Euclidean space, time and space are absolute, time may be modelled by a one-dimensional continuum, etc.

Allowing higher-level beliefs or aims into the hard core would allow for a more flexible account of mathematical progress. With the inclusion of broader goals, conjectures would then no longer be seen as the sole spurs to theory development. Although conjectures about mathematical objects still play a vital role in this century in carving out the future lines of attack – we only have to think of the developments prompted by Hilbert's twenty-three problems[5] – this class of aims used in the characterisation of research programmes must be extended to include such goals as that of, say, extending Kummer's theory to general algebraic number fields, or that of creating a non-commutative version of differential geometry. Not only will this allow us to account for the historical fact that not all mathematical research programmes are launched by the progressive understanding generated by improved attempts at solving a single conjecture, but it will also provide us with something that may be adhered to with tenacity, a quality Lakatos expected of his *hard core*, in that more general aims cannot be decisively established or refuted.

As an illustration of this last point let us compare two aims of importance to geometers, one at the level of conjecture, the other described more as a broad aim:

algebraic objects (e.g. groups), and topology itself can be used to explore algebraic entities, as is done in topological algebra.

[5] Of course, several of Hilbert's problems are expressed in terms of broad goals.

(1) To prove or disprove Poincaré's conjecture that all spaces homotopy equivalent to the 3-sphere are homeomorphic to the 3-sphere

(2) To develop a good account of three-dimensional manifolds.

Although it may be supposed that a solution to (1) will involve contributions to (2) and vice versa, it is possible that the conjecture will be decided one way or the other by an uninformative proof or an uninstructive counter-example, neither of which helps with the second aim. In the former case, we may end up with a computer proof of the sort used to prove the four-colour theorem which left many mathematicians with the feeling that unless a more conceptual proof is devised at some point, the conjecture was not a good one in the first place. For them, understanding is just as important as, or more important than, deciding the correctness of a result. While, in the latter case, if a counter-example is found, then topologists will still want to find the reasons for the counter-example's not being homeomorphic to the 3-sphere or to find the number of different homeomorphism classes.

William Thurston is one of the leaders of a programme which is working to achieve aim (2). He has proposed the following conjecture, one that subsumes Poincaré's:

The interior of every compact 3-manifold has a canonical decomposition into pieces which have geometric structures. (Thurston 1982: 357)

The whole point of a conjecture is to mark out the path to the achievement of some central aim, and as such belongs to the positive heuristic. As Thurston remarks:

just as Poincaré's conjecture, [it] is likely not to be resolved quickly, but I hope it will be a more productive guide to research on 3-manifolds than Poincaré's question has proven to be. (1982: 358)

Thurston also provides in this article a list of twenty-four 'questions and projects' on 3-manifolds and Kleinian groups which constitute targets for the programme.

If with Lakatos the components of a research programme were paired (hard core–negative heuristic, positive heuristic–protective belt), my suggestion would bring the hard core and positive heuristic closer thereby threatening to collapse the whole construction. Already in the case of natural science, Lakatos has an inkling of this problem when he declares that '[t]he demarcation between "hard core" and "heuristics" is frequently a matter of convention' (Lakatos 1978a: 181n.).[6] One way out of this problem

---

[6] According to Zahar's account the metaphysics of the hard core leads to certain prescriptions that may be translated into the meta-statements of the positive heuristic. Thus 'the heuristic may reflect

might be a classification of beliefs as means or ends. The hard core is composed of ends or aims, while the positive heuristic is composed of favoured means to achieve these ends.

Although we have already met with certain difficulties in carrying over the research programme structure to mathematics, let us continue with the comparison. Algebraic topology as a programme could be said to be born refuted given the early understanding that it would be impossible to distinguish between all the different homeomorphism classes of subsets of the plane. The solution to the question of the 'right' set of spaces for algebraic topology to classify is thus constrained by the desire for generality and the desire for completeness.[7] Claims about which topological objects ought to form the study of algebraic topology we might assign to the protective belt. If we find that algebra cannot make the necessary distinction between two such objects, the negative heuristic would advise us to consider altering these claims.

As for the positive heuristic, we can take this to include the plan to start with the classification of triangulable spaces and then to extend this to more general spaces; to engage with neighbouring branches; to relate invariants of geometric objects to the invariants of their underlying topological spaces; to introduce as much power into the algebra as possible without rendering the translated problems as difficult as the original topological problems. The positive heuristic of a programme must be allowed to evolve: it would be incorrect, for example, to say that Poincaré presaged the extension of homology theory to a wider class of spaces since the very notion of space underwent dramatic change in the years following his *Analysis Situs* papers, driven by considerations unrelated to his own.

The spectre of a problem we shall meet repeatedly has arisen: the delineation of a research programme is made difficult by the presence of mathematicians temporarily working on a problem with different motivations and from different backgrounds. Witness the example described by Saunders Mac Lane (1978), where five mathematicians produced four independent papers continuing an idea of Hopf's in precisely the same direction.

By 1942, it had long been known that the first homology group of a space was the Abelianisation of its fundamental group. What Hopf achieved in a paper published in that year was the discovery of a relationship between

certain aspects of the hard core; the distinction between hard core and heuristic is not as absolute as Lakatos imagined' (Zahar 1989: 22).

[7] A further consideration requires that the space of mappings between two objects be itself an object of the same kind.

the fundamental group and the *second* homology group. Within three years of its publication no less than four papers, including one by Hopf himself, pushed this idea further, thereby laying the foundation for homological algebra. The conditions for such a multiple discovery were created by the difficulties in communicating papers during the Second World War. Mac Lane, a co-author with Eilenberg of one of the four papers, mentions how he recognised the construction of Hopf's as the Schur multiplier of a group. He had become acquainted with this construction through his study of class field theory, the continuation of the algebraic number theory of Dedekind to be discussed below. In a second of the four papers, use was made of the universal covering space, an indication that Galois theory is lurking in the background. Further ideas are found in the other two papers. So here we can already see the difficulties in untangling the lines of development of a theory. We have five mathematicians with their different backgrounds and knowledge of a variety of areas of mathematics involved for a brief time in the attempt to achieve similar ends, yet making use of dissimilar means. To count these mathematicians as working within the same research programme would force us to include within the heuristic of that programme a wide variety of favoured means originating from a considerable part of the totality of mathematics. When mathematicians from a variety of backgrounds work temporarily on the same project, the correct attribution of the heuristic becomes a far from simple matter.

As for equivalents of Lakatos's three varieties of progress, we can still talk of developments being in the spirit of a programme and so of heuristical progress: progressive research should have a sense of purpose and employ suitable means to achieve its ends. If a programme generates new concepts, produces new proof techniques, elaborates on partially understood ideas thus enabling generalisation of results to a wider domain or a neighbouring field, then we could talk of theoretical progress. If a programme permits problems in other fields to be solved, extends our ability to calculate solutions of equations or investigate the nature of these solutions, or provides new applications in science or engineering, then we might say it constitutes empirical progress. Mathematicians are well aware of these varieties of progress:

Mathematics is good if it enriches the subject, if it opens up new vistas, if it solves old problems, if it fills gaps, fitting snugly and satisfyingly into what is already known, or if it forges new links between previously unconnected parts of the subject. It is bad if it is trivial, overelaborate, or lacks any definable mathematical purpose or direction. It is pure if its *methods* are pure – that is, if it doesn't cheat and tackle one problem while pretending to tackle another, and if there are no gaping holes in its logic. It is applied if it leads to useful insights outside mathematics. By these criteria, today's mathematics contains as high a proportion of good work as at any other period, and as any other area; and much of it manages to be both pure and applied at the same time. (Stewart 1987: 233)

The importance of a part of mathematics is something one can judge roughly by the amount of interaction it has with other parts of the subject . . . Hard core mathematics is, in some sense, the same as it has always been. It is concerned with problems that have arisen from the actual physical world and other problems inside mathematics having to do with numbers and basic calculations, solving equations. This has always been the main part of mathematics. Any development that sheds light on these topics is an important part of mathematics. (Atiyah 1984: 299)

Algebraic topology has shown all these varieties of progress. In terms of heuristical progress, homology theories were devised of greater applicability from cell complexes to general topological spaces, and ever-more sophisticated tools (e.g. squaring operations) have provided for finer classifications. Theoretical progress has been achieved by the development of homological algebra, category theory, and K-theory, which have had effects on other branches of mathematics far wider than could have been hoped by their discoverers, who certainly did not lack ambition, thus adding to the complex intertwining of modern mathematical theories. From Poincaré's qualitative analysis of the solution sets of differential equations, through Lefschetz's fixed point theorem and Morse's calculus of variations, to the use of homotopy classification of mappings in modern conformal field theories, algebraic topology has extended the understanding and solution of equations in mathematics and physics throughout its history and therefore may be deemed to have undergone empirical progress. One of the main reasons for the introduction of the methodology of scientific research programmes was to provide rules (if only retrospective ones) to determine the relative status of competing research programmes, and thus to class the refusal to acknowledge the degeneration of a research programme as irrational. Here the aim was to account for what was better about, say, Einsteinian physics compared to its Newtonian predecessor. One could compare this example of competition with the success of Eilenberg–Steenrod axiomatic algebraic topology over earlier formulations. What these axioms do is to define what it is to be a homology theory in terms of properties already known to hold for existing homology theories. It was a radical change less for the content of the theories, more as an example of a new way of doing mathematics: the use of axioms to define a class of processes that acts on one type of object to produce another type of object. Mathematics became for the first time explicitly *functorial*. This brings us to one of the central ideas of this chapter: rivalry in mathematics is less over the ground the competing theories are covering, and more concerned with high-level questions regarding the organisation of mathematics.

What of more direct confrontation of theories? Can we find examples of competing research programmes? Point-set (or general) topology might be thought of as a rival to algebraic topology, and yet, despite a certain overlap in their goals, it should surely be seen as aimed at answering different questions or as providing a useful, but distinct, set of conceptual tools for other branches. Algebraic topology has provided the different branches of geometry with a host of useful technical devices and given a huge impetus to algebra; point-set topology has been useful in, among other areas, functional analysis and Boolean algebra. Algebraic topology is concerned with very different aspects of a topological space, e.g., its 'holedness', whereas point-set topology explores aspects of separability. One can, however, debate their relative importance for mathematics as a whole. While discussing a recent development in model theory, Angus Macintyre commented that:

> [t]his can be construed as a long-overdue rapprochement between logic and <u>algebraic</u> topology (as opposed to set-theoretic topology). Kreisel had stressed the gigantic difference in importance (for the life of mathematics) between algebraic topology (coming from Poincaré) and set-theoretic topology. The latter is of course ubiquitous in routine arguments and formulations, but the former is almost unreasonably effective in advancing mathematical understanding. (Macintyre 1989: 366)

But this type of theory comparison bears little resemblance to those Lakatos described as deciding between rival scientific research programmes competing in similar observational realms.

We can, however, find examples of fairly direct competition within modern mathematics. Comparison of research programmes for Lakatos involved the notion of the degeneration of programmes. A programme is degenerating when all it can do to counter anomalies is to make ad hoc, or non-progressive, changes to its protective belt. We can find many examples of this phenomenon of stagnation in mathematics, but I shall confine myself here to just one.

One of the areas of mathematics that has been seen as unapproachable from a constructive perspective, despite the efforts of Errett Bishop, is point-set topology. Constructivists have been unable to treat topological spaces in their full generality, restricting themselves instead to metric spaces. But given that most of the results on topological spaces rely solely on the lattice structure of their open sets, a constructive counterpart can be defined in what is known as a *locale*, a type of lattice. In the words of Peter Johnstone, a mathematician working in this field:

> It is, I think, by now generally accepted that topos theory has something to say to constructive mathematics. But I believe that the contribution of topos theory should be more than merely providing models against which the constructivists can test their own (preconceived) ideas; it also has a role to play in suggesting what constructive mathematics ought to be – what results one should aim for, and even how one should try to prove them. In this latter sphere, topos theory has so far made little headway; even the message that constructive general topology ought to be about locales and not spaces, which has been broadcast loud and clear by topos theorists for some years now, has had little impact on any of the traditional schools of constructive mathematics. (Johnstone 1984: 84–5)

For constructivists to use toposes as models to test their ideas against is surely ad hoc, in the sense of being heuristically non-progressive, since these structures are not constructively defined. They are category theoretic structures that happen to model 'internally' higher-order intuitionistic logic. They were developed by category theorists from work done in algebraic geometry and point to the practical rather than philosophical importance of constructivism. The reliance on theory solely generated by other programmes is a sure sign of degeneracy.

Again the dispute is at a high level: the category theorist versus the constructivist. To examine this idea that rivalry is always at a high level, I shall now look at an example of rivalry at an apparently lower level by comparing the fates of two extensions of Kummer's algebraic number theory – Dedekind's versus Kronecker's – as seen by later mathematicians.[8]

8.3 DEDEKIND VERSUS KRONECKER

Recall from chapter 4 that Kummer created the notion of an ideal divisor in response to the anomaly that for certain number domains prime factorisation is not unique.[9] For instance, in $Z[\sqrt{-5}]$ we can factorise 6 in two ways: $(1 + \sqrt{-5}) \cdot (1 - \sqrt{-5}) = 6 = 2 \cdot 3$, such that the factors cannot be reduced further. Kummer's idea was to imagine that these factors could be reduced further into ideal divisors such that the two factorisations were alternative ways of grouping the ideal divisors. He then gave conditions

---

[8] The ideas of the Russian mathematician Zolotarev, who also worked in this area, could profitably be studied as a third rival.

[9] Harold Edwards has discussed Kummer's research, along with its development by Dedekind and Kronecker, in great detail in several books and articles. For Kummer's work see, for instance, Edwards (1977).

for divisibility by an ideal divisor in terms of the elements of the number domain.

Dedekind replaced Kummer's notion of an ideal divisor by the *ideal* of integers divisible by that divisor in his attempt to extend unique factorisation to general number fields. Kronecker achieved the same goal in a paper which appeared in 1882, despite the theory having been worked out many years before, in the same edition of *Crelle's Journal* as its rival the Dedekind–Weber paper. His major innovation was to investigate polynomials in a finite number of indeterminates with coefficients in the number field. He then defined the *content* of a polynomial as the greatest common divisor of its coefficients with a view to identifying associate polynomials, i.e., polynomials with the same content. Before this identification is made, however, he has the advantage of a simple notion of addition, subtraction and multiplication and, allowing rational functions, he is also able to divide.

Kummer showed clear partiality for one of the heirs to his theory over the other. According to Dedekind, Kummer did not want to know about his research, while admiring his former pupil's ideas. To help us appreciate the difference between the styles of these two protagonists it will pay us to jump forward several decades to consider a section of Weyl's book on algebraic number theory (Weyl 1940) entitled 'Our Disbelief in Ideals'. Here, Weyl presents two arguments against ideals. The first is that Dedekind's approach has the 'awkward consequences' that the ideal generated by a set of integers varies depending on the field in which it is embedded, so that, in particular, an ideal is prime only relative to a given field. Where Dedekind looks only at one field, Kronecker wants to talk in terms of concepts independent of the field under consideration. With Dedekind, if you wish to change the field you must intersect the ideals with a lower field or take the ideal they generate in a larger field. Divisors, on the other hand, are defined such that nothing changes if the field is extended:

It is a remote consequence of the theory that both requirements agree, while in Kronecker's theory the embedding field . . . is irrelevant for the definition. (Weyl 1940: 67–8)

The second argument arises when we wish to look beyond algebraic number fields with the recognition that in a ring of polynomials in more than one variable, while the ring already possesses unique factorisation, ideals do not. So, whereas the use of ideals restores unique factorisation in number fields, it destroy it in some rings. Thus:

[o]ur aim here is to secure the law of unique decomposition. With this sole purpose in mind we must reject Dedekind's notion of an ideal as a universal solution. (Weyl 1940: 38)

Differences between the two approaches emerge more markedly when we pass to fields of polynomials in several variables with complex coefficients. The algebraic geometer wishes to study manifolds determined as the zeroes of a set of algebraic equations, which may thus be equated with the ideal of polynomials everywhere zero on the manifold. He will thus want to make the distinction between the surface $f = 0$ and the surface $f^2 = 0$, and this ideal theory allows him to do. Weyl admits:

that polynomial ideals are a worthy subject of study – not, however, as a tool for the arithmetic of polynomials, but for their own sake, because algebraic manifolds of lower dimension deserve no less attention than algebraic surfaces. (Weyl 1940: 38)

He finishes up with the following compromise position:

In summarizing, one may venture to say that K[ronecker] is the more fundamental, D[edekind] the more complete theory; or that D is of higher importance to the geometer, who ought to be concerned about manifolds of every dimension, while K is more important to the arithmetician, whose chief concern (presuming he is old fashioned enough!) is the law of unique factorisation. (Weyl 1940: 70)

The use of the word 'disbelief' in Weyl's choice of title for the section in which these thoughts are discussed may seem a little curious in the context of an argument about the relative utility of divisors and ideals. Indeed Weyl says later in the book:

As both theories are actually equivalent one can dissent about questions of convenience only. To my judgment the odds are here definitely against Dedekind. His theory suffers from a certain lack of self-sufficiency, in so far as its proofs resort to indeterminates and pivot around the fundamental Lemma . . . tools which are native to Kronecker's set up, alien to Dedekind's. (Weyl 1940: 67)

Thus, we might say that Weyl is accusing Dedekind of being ad hoc in the sense of not being in line with the heuristics of his programme. However, this judgment he notes in an amendment to the book (1940: 223) is not fair to the Dedekindian approach as a whole in that Noether and Krull had in the intervening period worked out a way of avoiding this incongruous move.[10] Surely then the only way to account for Weyl's preference is on the basis on a deeper ideological split. We will not have far to look for

---

[10] As it happens, Weyl was being unfair to Dedekind as the latter had himself avoided resorting to indeterminates.

support for this claim after recalling that Weyl had strong constructivist leanings and so was in much closer accord with Kronecker's beliefs than with a philosophy that could be instrumental to the founding of set theory. As Weyl points out:

> Kronecker's criterion of divisibility is one decidable by finite means, while Dedekind's criterion refers to the infinite set of all possible integers. (1940: 67)

The differences between the two approaches reflect real philosophical divergences. Whereas divisors can be computed with, Dedekind did not have an algorithm for deciding whether a given element is in an ideal and this Kronecker and later mathematicians of a constructive bent saw as being 'needlessly nonconstructive' (Edwards 1989: 68).[11]

Having been commissioned to write a review of the state of research in algebraic number theory at the end of the nineteenth century, David Hilbert used the opportunity in his *Zahlbericht* to rethink the whole field, opting largely for the Dedekindian approach, which in this way became standard.[12] Certainly Dedekind's approach won out over the decades leading to van der Waerden's *Moderne Algebra*, but already by 1949 the *p*-adic approach of Kronecker's student Hensel was seen by many as a strong competitor:

> It seemed at first that the ideal-theoretic approach was superior to the divisor-theoretic, not only because it led to its goal more rapidly and with less effort, but also because of its usefulness in more advanced number theoretic research. For Hilbert and, after him, Furtwängler and Takagi succeeded in constructing on this foundation the imposing structure of class field theory, including the general reciprocity law for algebraic numbers, whereas on Hensel's side no such progress was recorded. More recently however, it turned out, first in the theory of quadratic forms and then especially in the theory of hypercomplex numbers (algebras), not only that the divisor-theoretic or valuation-theoretic approach is capable of expressing the arithmetic structural laws more simply and naturally, by making it possible to carry over the well-known connection between local and global relations from function theory to arithmetic, but also that the true significance of class field theory and the general reciprocity law of algebraic numbers are revealed only through this approach. Thus the scales now tip in favor of the divisor-theoretic approach. (Hasse 1980: vi)

---

[11] Note, however, that in a paper published in 1932, Weyl speaks well of ideals. The concept of an ideal is described as 'fundamental' (1932: 649) and as playing a 'dominant role' (1932: 650) in commutative and non-commutative algebra.

[12] Weber had presented Kronecker's version in his *Lehrbuch der Algebra* (1896). Why, then, Hilbert's choice? Should one point to an overlap in their methodological outlook, or might there be an element of contingency? For an account of the mathematical styles of Dedekind and Hilbert see chaps. 2 and 3 of Corry (1996).

Hensel's ideas were closely aligned with Kummer's divisor theory and were suggested by the analogy between algebraic number fields and algebraic function fields first explored, paradoxically, by Dedekind and Weber, as we saw in chapter 4. Their joint paper, however, borrowed elements from the divisor theoretic tradition. Triumph for Hensel, however, was not triumph for Kronecker:

There appears to have been a certain feeling of rivalry, both scientific and personal, between Dedekind and Kronecker during their life-time; this developed into a feud between their followers, which was carried on until the partisans of Dedekind, fighting under the banner of the 'purity of algebra', seemed to have won the field, and to have exterminated or converted their foes. Thus many of Kronecker's far-reaching ideas and fruitful results now lie buried in the impressive but seldom opened volumes of the Collected Works. While each line of Dedekind's XIth Supplement, in its three successive and increasingly 'pure' versions, has been scanned and analyzed, axiomatized and generalized, Kronecker's once famous *Grundzüge* are either forgotten, or are thought of merely as presenting an inferior (and less pure) method for achieving part of the same results, viz., the foundation of ideal-theory and of the theory of algebraic-number fields. In more recent years, it is true, the fashion has veered to a more multiplicative and less additive approach than Dedekind's, to an emphasis on valuations rather than ideals; but, while this trend has taken us back to Kronecker's most faithful disciple, Hensel, it has stopped short of the master himself . . . Now it is time for us to realize that, in his *Grundzüge*, Kronecker did not merely intend to give his own treatment of the basic problems of ideal-theory which form the main subject of Dedekind's life-work. His aim was a higher one. He was, in fact, attempting to describe and initiate a new branch of mathematics which would contain both number-theory and algebraic geometry as special cases. This grandiose conception has been allowed to fade out of our sight, partly because of the intrinsic difficulties of carrying it out, partly owing to historical accidents and to the temporary successes of the partisans of purity and of Dedekind. (Weil 1950: 90)

Weil then goes on to say that:

It will be the main purpose of this lecture to try to rescue it from oblivion, to revive it, and to describe the few modern results which may be considered as belonging to the Kroneckerian program. (1950: 90)

Weil's interest in this programme was caused not through any great concern for the constructive tradition *per se*, but rather through his belief in the mathematical power of Kronecker's approach, one that embraced far more than the constructivism for which he is often remembered. As it turned out he was proved to be fully justified in his beliefs when in the late 1950s and 1960s Grothendieck's massive reformulation of algebraic geometry subsumed Kronecker's programme. Weil modestly claims later that this revival in the fortunes of Kronecker's theory was not due to his own contribution, but that it had arisen from considerations other than those he had mentioned:

My foremost desire was to draw attention to the opportunity of studying algebraic geometry over a ring. Doubtless the natural development of the subject would have led there by itself; whatever the case, it has been satisfied to a large extent . . . above all by the theory of schemes devised by Grothendieck and developed by his students and successors. (Weil 1979: 576, my translation)

A hint of the constructivity of Kronecker still survives in that Grothendieck devised for his programme a class of categories, known as Grothendieck toposes, and their generalisation, the *elementary* toposes of Lawvere and Tierney, are models for higher-order constructive logic. But what the many mathematicians, including Langlands, who have favoured Kronecker over Dedekind have realised is the greater vision of the former's idea. The point here is that the concerns of algebraic geometers have proved to include and surpass most of what is important for mathematics in the idea of constructivity. This view is also held by the logician Gonzalo Reyes:

work on 'foundations' has been concentrated for too long on the field of Analysis and the dialectic contradiction of 'constructive' versus 'non-constructive.' It seems to me that it is high time to look for 'foundational' fields elsewhere and Algebraic Geometry with a much richer dialectics seems as good a choice as any. And who knows, it may even be the Ariadne's thread of the 'labyrinth of the continuum.' (Reyes 1980: 250)

As regards the rivalry of the two programmes, we have seen that both have succeeded as successors to Kummer's programme in their different ways. The language of ideal theory occurs throughout mathematics and yet, in a sense, Kronecker's approach was the more fundamental. It aimed more directly at the very heart of the seemingly inexhausible analogy that exists between number and function. This situation where two theories appear to converge on a common problem yet later turn out to proceed in different directions is, I suggest, very common in the history of mathematics, more so at any rate than rival scientific theories accounting for the same experimental data.

We may see two consequences following from this short case study. First, it implies the necessity of a many-layered account of mathematics, where theories can succeed at different levels. Second, as mathematicians have in some sense a little more room to manoeuvre than scientists, there is less chance for them to be disputing precisely the same territory.

8.4 OTHER ATTEMPTS TO CONSTRUCT A METHODOLOGY OF MATHEMATICAL RESEARCH PROGRAMMES

Only a few attempts have been made to extend the methodology of scientific research programmes to mathematics. These include those outlined
in articles by Michael Hallett and Giulio Giorello, about the first of which
(Hallett 1979) I shall not have much to say. Here Hallett concentrates on
the issue of progress, proposing that to say that one theory has progressed
over another if it satisfies *Hilbert's criterion*. This states that:

the setting up of a new mathematical theory T_{m+1} constitutes progress with respect
to its predecessor T_m if T_{m+1} is used in the solution of at least one problem P
which T_m did not solve, provided that P is not of T_{m+1}'s own making and that the
statement solving P was not used in the construction of T_{m+1}. (Hallett 1979: 10)

This is clearly closely related to Lakatos's empirical progress with Zahar's
modification. Hallett shows quite convincingly that point-set topology
exhibited this variety of progress through the early years of this century and
provides support for the notion of a methodology of mathematics research
programmes.

Turning now to Giorello (1980) we find the author alleging that:

In our opinion, (in the context of mathematical discovery) *Lakatos' perspective
provides more adequate tools than Kuhn's.* (1980: 118, author's emphasis)

This remark is made on the basis that:

in the growth of knowledge, *concatenations* between scientific achievements are
observed which no typology should ignore, but strangely enough, the author of
The Structure of Scientific Revolutions seems to pay little attention to this. (1980:
118, author's emphasis)

This is perhaps some truth to this in the light of the lack of dialectical
awareness shown by Kuhn, as I discussed in chapter 1.

Giorello takes the story of Riemann's development of the theory of functions of a complex variable as a case study to back his claim. He isolates
three principles in Riemann's programme: (a) the Dirichlet principle; (b)
the definition of mathematical entities by means of their behaviour in the
infinitely small; and (c) some 'metaphysical' principles taken from Herbart
such as that 'therapy' was necessary to counteract the 'prejudices inadver-
tently transmitted through language'.

There is no attempt by Giorello in this article to outline the range of
Riemann's work and so to enable us to see the roles these three principles
play for Riemann. To me it seems a curious decision to want to lump

them together. The Dirichlet principle is a technical tool which provides the means for obtaining a real harmonic function, u, on a domain Ω coinciding with a given continuous function on the boundary of Ω by finding an appropriate u that minimises a certain integral $D(u)$. This is justified by the observation that $D(u)$ is always non-negative.

As for (b), this may be appropriate to describe Riemann's work in differential geometry, and accords with Weyl's description of Riemann's quest 'to understand the world from its behavior in the infinitely small' (Laugwitz 1999: 307). As for complex function theory, it better accords with Weierstrass's approach. Weierstrass, we may recall, understood these functions in terms of their series expansions about different points; Riemann on the other hand, adopted a global approach by looking for suitable spaces on which they could be defined.

Finally, (c) involves general philosophical ideas held specifically by Riemann himself and not by his followers. Such ideas have commonly been held by very imaginative mathematicians and tell us much about the creative thinking behind their work, yet to include it in the hard core would suggest that these beliefs belonged to anyone working on the programme.

Giorello describes the successive attempts to establish the limits of the domain of validity of Dirichlet's principle as continuations of Riemann's programme, thus sees (a) as belonging to the community of mathematicians working on the programme. But surely this is to confuse the individual with the community faces of mathematics. It is extremely unlikely that Riemann's successors adhered to his Herbartian beliefs or even necessarily agreed with his views on the infinitely small. Why not rather define the programme as the attempt to establish a coherent theory of complex functions in a global style, these later efforts being attempts to deal with anomalies in the programme? A separate venture would be to explore the personal heuristical unity behind Riemann's work.

In his (1991) Koetsier criticises Giorello for a rather loose application of the terminology of a research programme but, as we can see from the following statement, Giorello has realised that the research programme language will not translate directly over into mathematics:

Do we hold then that one of the more distinctive characteristics of the methodology of scientific research programmes – the distinction between the metaphysical hard core and the protective belt, in characterizing the heuristics – cannot be applied to case studies such as the one under consideration, or that it should be abandoned? As an absolute scheme, this distinction, in our opinion, leads to considerable difficulties. However, if it is assumed as a working hypothesis, in a flexible, non-schematic way, it will allow us to characterize assertions such as those of type (c), in our example as at once 'metaphysical' and 'auxiliary' according to our interests in the reconstruction of the research program. (1980: 130–1)

I have, however, based the present chapter on the assumption that is only by a more rigorous application of Lakatos's ideas that we will be able to extract what is best about them and so produce a clearer picture of mathematical progress.

Koetsier claims that:

the heart of Riemann's scientific enterprise is first of all the conviction that a very general, global approach to the theory of complex functions of a complex variable is possible. Secondly, the theory of complex functions was linked to geometry (conformal mappings) on the one hand and to potential theory on the other. (1991: 143)

While the accuracy of this characterisation of the work of Riemann and his followers as a programme is open to debate, it is noticeable that Koetsier here is talking effectively in terms of ends and means, and not axioms and theorems. Let us now turn our attention to Koetsier's replacement for the methodology of mathematical research programmes.

8.5 KOETSIER AND THE METHODOLOGY OF MATHEMATICAL RESEARCH TRADITIONS

In his (1991), having offered criticisms of attempts to transfer the methodology of scientific research programmes to mathematics, Koetsier tries to do better with Laudan's notion of a research tradition. This he defines as follows:

A *mathematical research tradition* is a group research activity, historically identifiable (in a certain period), characterized by common general assumptions (in the form of e.g. definitions, axioms) about the entities that are being studied in a *fundamental mathematical domain*, and it involves assumptions about the appropriate methods to *prove* properties of those entities.
  A *fundamental mathematical domain* consists of the most general mathematical entities that play a role in the mathematics of the period. (1991: 151)

Koetsier offers the following 'tentative' notion of mathematical rationality:

A mathematical research project or research tradition *progresses heuristically* if it produces conjectures (theorem candidates) of weight. Apart from heuristic progress there is *absolute progress*, if the project or the tradition succeeds in proving or refuting the conjectures resulting from its heuristic progress, it progresses absolutely. But we have also absolute progress, if the project or tradition succeeds in proving (or refuting) conjectures or theorems produced by a competing project or tradition . . .

The preference of a rational mathematical community for a research project or a research tradition is proportional to its expected progress. (1991: 159)

The first doubt I should like to express about this picture is as to whether progress should be judged purely in terms of the theorems or conjectures generated by a tradition or project. Other candidates for signs of progress include the reorganisation of existing bodies of work and the production of new techniques to solve problems which need not be theorems, for example, enabling one to solve a new class of differential equations. It is all too easy to forget that one of the main purposes of elaborating formal concepts is to provide methods of calculation – recall the quotation from Atiyah above.

The notion of the importance or 'weight' mathematicians give to some theorems over others needs analysis and Koetsier provides this by listing nine factors. When we come to look closely at these we find that several of these factors are expressed in terms of theories. Thus it begins to look as though Koetsier has realised that there is more to mathematics than its theorems. However, when we turn to his definition we find Koetsier claiming that a theory belonging to a tradition is:

a set of connected mathematical statements that are considered to be true of (a subdomain of) the fundamental domain of that tradition encompassing at least one theorem, i.e. a proven conjecture. (1991: 153)

Once again logical positivist doctrine has left its mark and has restricted us to a set of statements.

Next, there follows the problem of deciding the best scale with which to classify units of research. Given the dramatic rise in the number of working mathematicians since the Renaissance, might the most appropriate size not vary through the centuries? Koetsier's classification of mathematical theories as belonging to research traditions seems to me to be increasingly inappropriate as we move forward to the present day. As he says about analysis (1991: 199), this branch was not all of mathematics in the nineteenth century, but was of such importance that it affected the whole community. This is more or less correct. The adherence of their aging teachers to the view that analysis was the pinnacle of mathematics drove the future founders of *Bourbaki* to Göttingen and Hamburg to find out all about the new German abstract algebra, the intervening generation having been wasted in the trenches of the Western Front. But what of twentieth-century research traditions? Koetsier tells us only of the 'structuralist' tradition, the one promoted by *Bourbaki*, whose fundamental domain 'consists of the totality of all mathematical structures, and its methodology of proof is the modern formalist axiomatic method' (1991: 152). This tradition so described includes the majority of modern mathematics, and so the term 'research project' is then left to do an immense amount of work. In effect it has to describe all the developments that have occurred since, say, the 1930s.[13]

Surely we should aim for a resolution finer than one which takes modern algebraic topology as a mere project. After all, *Mathematical Reviews* classifies research in this branch under roughly a hundred sub-headings, all of which are active.[14] Seen as a whole, algebraic topology deserves to be classified on a level different from that of one of its research problems, for instance, the attempt to discover the higher homotopy groups of the spheres, or the theory of knots. These latter would seem to merit the name *project*, while the former, certainly by the Steenrod–Eilenberg axiomatisation, deserve the name *programme*. Why not then allow for three levels – tradition, programme, and project – each of which may fade away through lack of results or be superseded or swallowed up by a rival, the latter two also being able to spur new developments in the next higher level? In Kronecker we can see elements of an algorithmic tradition, a programme to unify arithmetic and algebraic geometry, and a project to better ground Kummer's ideal numbers. We should not access the degree of success of his work as a monolithic block.

Category theory, it could be argued, began as a project to study continuous mappings within the programme of algebraic topology, has since become a programme in its own right, and is challenging set theory to become the language of the dominant tradition. It claims to provide new means better adjusted to the needs of the structuralist research tradition.

In the twentieth century we see the increasing take-over of the structuralist tradition,[15] where branches, which had been kept artificially isolated from each other, were allowed to interact, often through projects starting within a particular programme weaving through other programmes. K-theory is a perfect illustration of this in that it originated in a project of Grothendieck and others within topology to classify vector bundles over a topological space and has since made contributions to arithmetic, algebra, and functional analysis. As such it has played a crucial role in the construction of areas of mathematics currently absorbing the interest of many mathematical

---

[13] Cf. Lakatos's remark: 'Even science as a whole can be regarded as a huge research programme . . . But what I have primarily in mind is not science as a whole, but rather *particular* research programmes' (Lakatos 1978a: 47).

[14] Papers are often classified under several subheadings.

[15] See Corry (1996) for an account of the rise of the structuralist approach. Corry makes the important point that the rise of structuralism should not be confused with the rise of axiomatisation (1996: 53 and n.) and devotes the second part of the book to a discussion of various attempts to capture the notion of structure.

physicists. A typical research programme today, such as the Langlands programme, a non-commutative generalisation of class field theory, is a far better integrated affair than those of the pre-war era.

This points to a difference between mathematics and physics: while scientific research programmes can be grouped into higher-level traditions characterised by a guiding metaphysics, for instance, an adherence to atomism or to a field theoretic outlook, it is often best to focus on the level of programmes or projects to see where the battles are being fought out. Mathematics appears to have an extra degree of freedom at this level which makes it improbable that projects, and even programmes, will be in direct competition for the precisely the same territory. Point-set topology and algebraic topology, we have said, study similar objects but in different ways, to the benefit of each other. As a consequence, the outcome of struggles between rivals is less often one of outright victory. To admit this extra degree of freedom is by no means to allow that anything goes.

Traditions spawn projects which may become programmes in their own right, may then link up in tandem with other programmes, possibly straddling different traditions. Riemann's initial account of connectivity and Betti's improvements were within the tradition of analysis as represented by Cauchy, the measuring being done only in terms of natural numbers. Poincaré's work in algebraic topology began as a project to help provide him with tools to study differential equations qualitatively, the spaces belonging to the older idea of a manifold taken from differential geometry, while the algebra was done in terms of the linear algebra of the time. Algebraic topology became a programme in the Princeton of the 1920s. Spaces had by then been generalised to the point-set variety founded by Hausdorff among others, while later Noether's rings and modules transplanted the matrices of the old linear algebra.

An idea of the complex shifting of levels comes from Atiyah (1974: 213) in which the author describes how when faced by a problem a mathematician may devise a 'trick' to help him solve it. Further analysis reveals to him how this trick works and allows him to develop a 'technique'. If the technique is generalisable it in turn may become a 'method', which further developed becomes a theory.

The greatest problem with the attempt to transfer pictures of scientific theoretical progress, developed often from historical research of the science of earlier times, to the mathematics of the twentieth century is the rapidity with which ideas came to be passed from hand to hand. The language of research programmes and projects as clearly delineated entities seems better suited to describe work conducted in the comparatively leisurely atmosphere of earlier centuries. Recall from above the speed with which the ideas in Hopf's 1942 paper were developed by himself and four other mathematicians, each from a different background. It is understandable then why mathematicians presenting a synoptic view of progress in a particular theory relate the history of ideas rather than the history of their discoverers.

We can even find mathematicians offering historiographical advice. For instance, Weil points out that:

In Bourbaki's historical note on the calculus, it is said that the history of mathematics should proceed in the same way as the musical analysis of a symphony. There are a number of themes. You can more or less see when a given theme occurs for the first time. Then it gets mixed up with the other themes, and the art of the composer consists in handling them all simultaneously. Sometimes the violin plays one theme, the flute plays another, then they exchange, and this goes on.

The history of mathematics is just the same. You have a number of themes; for instance, the zeta-function; you can state exactly when and where this one started, namely with Euler in the years 1730 to 1750 . . . Then it goes on and eventually gets inextricably mixed up with the other themes. It would take a long volume to disentangle the whole story. (Weil 1974: 291)

This technique of writing a history of mathematics was adopted by Dieudonné in his 'The Historical Development of Algebraic Geometry' (Dieudonné 1972), where the themes he selects there are: (A) Classification, (B) Transformation, (C) Infinitely near points, (D) Extending the scalars, (E) Extending the space, (F) Analysis and topology in algebraic geometry and (G) Commutative algebra and algebraic geometry. Notice that these themes operate at the different hierarchical levels I have been discussing and that they reveal the extent of horizontal transfer within levels. Mathematics starts to look like a tangled net and it becomes much less appropriate for a historian to offer a sequential narrative.

8.6 MADDY'S NATURALISM

At about the same time as I was revising Lakatos's conception of research programmes in terms of aims and means, Penelope Maddy was led through her thorough understanding of the recent history of set theory to do something very similar in the form of a means–end analysis. Her analysis of set theory given in (1997) and elsewhere is a fascinating illustration of her thesis that we should see mathematical developments not as settled on philosophical grounds, but as driven by the desire to satisfy mathematical goals.

While I thoroughly recommend her account of the reasons philosophers of mathematics should become naturalistic methodologists, I think it is still worth drawing attention to some slight differences of orientation between us. Possibly these arise merely from the fact that we study different areas of mathematics. In this regard, two features of contemporary set theory are quite prominent. First, the number of practitioners is fairly small, aiding the achievement of a firm consensus on most matters. Second, the programme has remained far more isolated from other branches than is the case with most programmes. The first of these points may lie behind a decision not to question the rationality of the goals of community (see, e.g., 1997: 198). In fields where there is more dispute, it is not just the means but the whole hierarchy of goals that comes to be questioned. In fact, means and goals are very intimately wrapped up with each other. How the questions which are real to a community (Jardine 2000), i.e., ones that mathematicians have an idea how to get started on, come to be introduced and sustained is something the methodologist must address.

Relating to the second feature noted above, set theory's isolation, when you dip into core mathematical activity you find an enormous diversity of means and ends jostling together. This flexibility is seen to be required to broaden mathematicians' understandings of a particular type of object, and as such is also important for the health of a whole branch of mathematics. About algebraic number theory Serge Lang could say:

If there is one moral which deserves emphasis, however, it is that no one piece of insight which has been evolved since the beginning of the subject has ever been 'superseded' by subsequent pieces of insight. They may have moved through various stages of fashionability, and various authors may have claimed to give so-called 'modern' treatments. You should be warned that acquaintance with only one of the approaches will deprive you of techniques and understandings reflected by the other approaches, and you should not interpret my choosing one method as anything but a means of making easily available an exposition which had fallen out of fashion for twenty years. (Lang 1970: 176)

This is extremely important. One imagines a new approach will come along and sweep all before it, but typically it only partial captures the situation. As Lakatos told us, we can never be certain that a research programme won't be profitably revived. One might have imagined, for instance, that the modern definition of a function as a set of ordered pairs was adequate to all there is to say about functions, and that the older Eulerian representation of functions by formulas was buried. Not so. Recently, for instance, with the computer assisted study of solutions to partial differential equations, numerical analysis has been joined by symbolic computation:

This book is an introduction to new computational methods in the theory of linear PDE [partial differential equations – DC]. To explain the terminology, the 'old' methods, well predating the advent of computers, are concerned with approximate *numerical* solutions based on difference approximations to differential equations. The underlying mathematical point of view here is that a function is given by a table of its values. In contrast, the new computational methods forming the subject of this book are *symbolic*, i.e., based on the idea that a function is best given by a formula, e.g., as an explicit polynomial or power series. (Kapranov 2001: 481)

In sum, my worry is that a rational reconstruction may be used to stifle debate. In 1910, we could easily have presented Dedekind's means of resolving the problem of prime factorisation as having successfully met 'the' goal, and then taken Kronecker's ideas as not quite up to the mark. But as we have seen elements of these ideas resurfaced very strongly in the mid-twentieth century. This is not to say that no stretches of mathematical may be described fairly straightforwardly in means–ends terms, just that we must not overrationalise our accounts, as Lakatos himself was wont to do, and we need to keep an eye on apparently failed programmes.

8.7 CONCLUSION

I have attempted to demonstrate that any model of the development of modern mathematics will requires more sophistication than a simple transfer of Lakatos's methodology of research programmes from science to mathematics. The model I have proposed is more sophisticated, but may well turn out to have flaws. It would be extremely useful to make a study of the notions of progressiveness as presented by mathematicians working in a range of eras and in a range of fields, and to compare these with the historical record. I shall look in greater detail at contemporary ideas of progressiveness in chapter 9.

I have stressed in the introduction that at present there are far too few studies of modern mathematics. I suspect that when at last we come to explore the development of mathematics in the twentieth century in a sufficiently detailed and historically sensitive way, what will probably emerge is a complex story, similar in many ways to those produced by the historians and sociologists of modern physics. In the account of physics presented by Peter Galison in his *How Experiments End* (Galison 1987) the author enters into a discussion (chapter 5) of how long-term, middle-term, and short-term constraints govern experimentalists' beliefs and actions. These may possibly be thought of as finding their parallels in commitments to beliefs involved in traditions, programmes, and projects. Coincidently, Galison talks of *programmatic goals* when he discusses the beliefs to which a scientist is committed due to middle-term constraints (1987: 249).

For those philosophers who reject the idea of mathematics as comprising a collection of statements, changes made to their picture of the development of mathematics must have an impact on their philosophy. The danger of allocating insufficient attention to the twentieth century is that we shall then be forced to extrapolate from our knowledge of earlier times. However, mathematicians through this century have increasingly seen fit to relate apparently unconnected theories. Indeed, it is the success of the structuralist tradition to have allowed for an accelerated interweaving of research. The rise of this tradition has provided a convenient setting for the employment of analogy, ever a potent heuristical tool, which has resulted in the quickening pace of transfer of ideas between fields. Thus, mathematics today progresses in ways in which it could not have progressed a hundred years ago. Pursuing the idea of mathematics as the 'science of analogy' should prove fruitful for philosophers.

While it is clear from what I have just said that I believe it will remain a useful strategy for those working in the history of mathematics to keep an eye open on developments in the history of science, we must of course expect differences to emerge. One I have proposed in the course of this chapter is that rivalry between competing researchers tends to involve high-level issues. In physics, the distribution of cases of decisive rivalry would seem to occur at lower levels. It remains to be seen whether this turns out to be a general principle.

Now let us see what we can make of differences in mathematicians' conceptions of the most important factors constituting progress.

CHAPTER 9

The importance of mathematical conceptualisation

A mathematician, like a painter or a poet, is a maker of patterns. If his patterns are more permanent than theirs, it is because they are made with *ideas*.   (Hardy 1940: 24)

All these difficulties are but consequences of our refusal to see that mathematics cannot be defined without acknowledging its most obvious feature: namely, that it is interesting.   (Polanyi 1958: 188)

9.1 VALUES IN MATHEMATICS

As with any academic community, mathematicians must devote a significant part of their time to promoting their research activities. This occurs both *externally*, with a view to improving the standing of mathematics relative to other disciplines, and *internally*, with a view to establishing the importance of specific research programmes. What is very noticeable when one encounters such promotion exercises is the enormous variety of qualities alluded to and the differences in emphasis placed on these qualities. Even if, as seems to be the case, a considerable consensus has persistently taken certain moments in the history of mathematics to be pivotal, importance does appear to be a time-dependent notion. Something vitally important for one generation may not seem quite so crucial for the next. But, alongside these rather predictable variations between generations of mathematicians, one also finds considerable dissimilarities between mathematicians of any given era. We would expect, then, that an exploration of contrasting opinions about what constitutes an important advance would reveal much about the evolution of competing images of mathematics and the tensions existing between them.

A first step is to glean what we can from mathematicians who meet with the question of importance in their roles as researchers, teachers, referees, textbook writers, grant body panellists and doctoral supervisors. As I mentioned above, what we quickly discover is that there are many criteria for judging the importance of a development and that there are differences among mathematicians as to the value they place on success according to these various criteria. For the purposes of this chapter, I propose to arrange these criteria, several of which we matched to Lakatos's notions of progressiveness in chapter 8, into five broad categories:

(1) When a development allows new calculations to be performed in an existing problem domain, possibly leading to the solution of old conjectures.

(2) When a development forges a connection between already existing domains, allowing the transfer of results and techniques between them.

(3) When a development provides a new way of organising results within existing domains, leading perhaps to a clarification or even a redrafting of domain boundaries.

(4) When a development opens up the prospect of new conceptually motivated domains.

(5) When a development reasonably directly leads to successful applications outside of mathematics.

Naturally, some developments may be rated highly according to several or perhaps all of the categories. In particular, it may happen that a reformulation of a body of existing theory or a unification of existing theories leads to new results in an already existing domain, but also points the way forward to a new area. Of course, there is a fine line between clarifying the boundaries of an old domain and extending beyond them into a new domain, but there are cases which are clearly on one side or the other.

What needs to be brought under close scrutiny is the tacit weighting given by the mathematical community to these different criteria. My perception is that, very reasonably, if a development is seen either to be doing well or to have the potential to do well according to the majority of the criteria, then interest is guaranteed. Take, for example, the Atiyah–Singer index theorem, a formula demonstrated in the early 1960s which links analytic information concerning an elliptic differential operator with topological information on an associated vector bundle, thereby relating important constructions in algebraic topology to the domain of partial differential equations. This immediately scored extremely highly on (1) and (2), highly on (3), and had the potential to score well on (5). Sure enough, the theorem later found its uses in quantum field theory.

A problem arises, however, when a development appears to do well on one front, but poorly on the others. An example here is the computer assisted proof of the four-colour theorem. Few now deny that the theorem is true, or that the various computer proofs warrant our belief in it, and yet a

widespread feeling persists that unless there is some more conceptual success, for example, by linking the theorem to other branches in illuminating
ways,[1] then little has been achieved. What interests me more, however, are
situations where the conceptualists are in the minority position. We can
express their concern as follows: despite the greater difficulty of scoring
according to categories (3) and (4), success here is not in general given
sufficient weighting. Their worry is, in other words, that the conceptual
aspect of mathematical activity is on occasions undervalued in that the
acquisition of results is favoured over the reorganisation and elaboration of
concepts.[2]

To be what I have called a conceptualist is not indicative of any specific
view as to the ultimate goal of mathematics. As a conceptualist, you may
well believe that the proper organisation of mathematical ideas is an end in
itself, but equally it could be that you see it as the most appropriate way of
providing tools to model the natural world, which you view as the essential
purpose of mathematics.

Complaints of a lack of conceptual appreciation are not hard to find. For
instance, although his work is generally considered to be extremely important, Mikhael Gromov considers that a book of his containing fundamental
insights on partial differential equations

is practically ignored because it is too conceptual. (Berger 2000: 187)

Now, it is no easy business defining what one means by the term *conceptual*.
One radical position, represented by a book such as *Conceptual Mathematics*
(Lawvere and Schanuel 1997), sees category theory as providing much of the
answer. Where set theory picks up on a few of our everyday structural concepts (collection, membership, union, etc.), category theory does so more
extensively in such a way that its concepts can be found in a multiplicity
of contexts. Without wishing to take sides here, I think we can say that
the conceptual is usually expressible in terms of broad principles. A nice
example of this comes in the form of harmonic analysis, which is based on
the idea, whose scope has been shown by George Mackey (1992) to be immense, that many kinds of entity become easier to handle by decomposing
them into components belonging to spaces invariant under specified symmetries. In the case of Gromov's book, on the other hand, the conceptual
core is expressed in terms of the *h-principle*, which holds, roughly speaking, that, in many geometric situations, obstructions to the construction
of solutions to partial differential equations arise only from topology.

---

[1] This is being done. See Thomas (1998).   [2] This imbalance is also noted by Laugwitz (1999: 22).

In a fascinating paper, which provides an excellent counterpoint to this one, the mathematician Timothy Gowers (2000a) stands up for the kind of mathematics which earned him a Fields's Medal. To some the field of what he refers to as 'combinatorics' appears as a collection of wholly unrelated problems, each requiring some clever trick to solve it, while fields such as algebraic number theory contain many general unified results. But by contrasting the problem solving to the theory building components of mathematical activity and identifying the field of combinatorics as one where the former prevails, Gowers does not mean to suggest that there is no common ground between the ways of arriving at results in combinatorics. Instead, he notes that the solution of combinatorial problems often leads to the production of 'somewhat vague general statements' (Gowers 2000a: 72) which then open up other problems for solution.[3] It would be interesting to observe the extent to which these implicit general principles can evolve to become explicit unifying theories.

To my mind the most straightforward access we can gain to these issues is via a case study analysis. What we require then is an example of a development whose fate hangs or has hung in the balance. In this respect an account of quantum groups, fascinating though this would be for our understanding of mathematical physics in the late twentieth century, will not fit the bill. Given the centrality of Lie groups and Lie algebras to mathematical physics, the pleasantly surprising discovery that 'quantum' deformations of examples of the latter exist was never going to be seen otherwise than as an important breakthrough. What we need to observe is an idea which, it is claimed by some, has suffered neglect because of a lack of immediate success in the more 'practical' categories, (1) and (5) of those outlined above.

The case I have chosen to treat in this chapter concerns the question as to whether the group concept should be extended to, or even subsumed under, the groupoid concept. Over a period stretching from at least as long ago as the early nineteenth century the group concept has emerged as the standard way to measure the degree of invariance of an object under some collection of transformations.[4] The informal ideas codified by the group axioms, an axiomatisation which even Lakatos (1978b: 36) thought unlikely to be challenged, relate to the composition of reversible processes revealing the symmetry of a mathematical entity. Two early manifestations of groups were

---

[3] For example: 'if one is trying to maximize the size of some structure under certain constraints, and if the constraints seem to force the extremal examples to be spread about in a uniform sort of way, then choosing an example randomly is likely to give a good answer' (Gowers 2000a: 69).

[4] See Wussing (1984).

as the permutations of the roots of a polynomial, later reinterpreted as the automorphisms of the algebraic number field containing its roots, in Galois theory, and as the structure-preserving automorphisms of a geometric space in the Erlanger Programme. Intriguingly, it now appears that there is a challenger on the scene. In some situations, it is argued, groupoids are better suited to extracting the vital symmetries.[5] And yet there has been a perception among their supporters, who include some very illustrious names, of an unwarranted resistance in some quarters to their use, which is only now beginning to decline.

My claim is that, although groupoids did well at reformulating old domains and pointing to new areas for exploration, they suffered from leading to too little in the way of new techniques for solving old, circumscribed problems. Thus, their early adoption required an inclination towards being conceptually adventurous. However, now that programmes using groupoids have becoming established, researchers can use them to work with more of an air of what we might call 'normal mathematics' within these programmes.

9.2 WHAT IS A GROUPOID?

When promoting a mathematical concept, it is never a bad idea to think up an illustration from everyday life. Ronald Brown (1999: 4), a leading researcher in groupoid theory, has provided us with a good example by considering possible car journeys between cities of the United Kingdom. Now, one approach to capturing the topology of the British road system is to list the journeys one can make beginning and ending in Bangor, the Welsh town where Brown's university is located. This possesses the advantage that the members of the list form a group under the obvious composition of trips, where the act of remaining in Bangor constitutes the group's identity element.[6] However, for a country so dominated by its capital city, it might appear a little strange to privilege Bangor and the act of staying put there. Each city might be thought to deserve equal treatment.

---

[5] If they succeed, then my account of the deficiencies in Lakatos's philosophy of mathematics discussed in chapter 7 will be supported. Lakatos's belief that 'elementary group theory is scarcely in any danger [of heuristic refutation]' arises from his idea that 'the original informal theories have been so radically replaced by the axiomatic theory' (Lakatos 1978b: 36). But extended informal notions of symmetry which arise from working with axiomatised theories elsewhere in mathematics may provide such a refutation.

[6] Note that trips are being considered here only 'up to homotopy'. In particular, taking a trip and then retracing one's steps is to be equated with staying at home.

Pleasant as it is to remain in Bangor, staying put in London should surely be seen as another identity element. Moreover, if you want to know about trips from London to Birmingham, it would seem perverse to have to sift through the set of round trips from Bangor which pass through London and then Birmingham, even if all you need to know is contained therein. And if ferry journeys are excluded, this method is perfectly hopeless for finding out about trips out of Belfast. More reasonable then to list all trips between any pair of cities, where ordered pairs of trips can be composed if the destination of the first trip matches the starting point of the second. Something group-like remains but with only a *partial* composition. On this basis Brown can claim that:

[t]his naïve viewpoint gives rise to the heretical suggestion that the natural concept is that of groupoid rather than group. (Brown 1999: 4)

As the mention of heresy indicates, the suggestion is far from universally accepted within the mathematical community. We read that Alain Connes, the French Fields's medallist, considers that 'it is fashionable among mathematicians to *despise* groupoids and to consider that only groups have authentic mathematical status, probably because of the pejorative suffix oid' (1994: 6–7, my emphasis). This explanation of the origins of such a strong sentiment may seem implausible, but there can be little doubt that the climate towards groupoids has not been exactly favourable.

Brown reproduces a passage from a letter sent to him by Grothendieck in 1985:

The idea of making systematic use of groupoids . . . , however evident as it may look today, is to be seen as a significant conceptual advance, which has spread into the most manifold areas of mathematics . . . In my own work in algebraic geometry, I have made extensive use of groupoids. (Quoted in Brown 1999: 7)

One might have expected that eleven years later the matter would have been settled, the 'evident' idea would have spread, but according to Alan Weinstein, a noted geometer, by 1996 the message had still not got through:

Mathematicians tend to think of the notion of symmetry as being virtually synonymous with the theory of *groups* . . . In fact, though groups are indeed sufficient to characterize homogeneous structures, there are plenty of objects which exhibit what we clearly recognize as symmetry, but which admit few or no nontrivial automorphisms. It turns out that the symmetry, and hence much of the structure, of such objects can be characterized algebraically if we use *groupoids* and not just groups. (Weinstein 1996: 744)

To counteract resistance to their use we find that three articles have been written and two Internet websites constructed with a view to their promotion. Such explicit promotion is quite unusual, although mathematicians are aware of the need to market their wares. In a humorous subsection of his book, entitled 'Commercial break', the algebraic geometer Miles Reid tells us that:

Complex curves (= compact Riemann surfaces) appear across a whole spectrum of maths problems, from Diophantine arithmetic through complex function theory and low dimensional topology to differential equations of math physics. So go out and buy a complex curve today. (Reid 1988: 45)

Now, this book is aimed at undergraduates – complex curves have not stood in need of any PR campaign for the purposes of recommending them to professionals for many decades. In fact, it would be hard to count yourself a professional mathematician without agreeing that complex curves are a good thing, even if your research never brings you particularly close to them. Groupoids, on the other hand, despite generating sufficient interest for an annual 'Groupoid Fest' to be held in their honour, still require some salesmanship.

Two of the promotional articles are due to Brown (1987, 1999), the first appearing in the long-established *Bulletin of the London Mathematical Society*, which publishes research and expository articles, while the second forms the opening article of the first issue of a new journal *Homology, Homotopy and Applications*. The other article, Weinstein (1996), appears in the *Notices of the American Mathematical Society*, a more informal journal, which includes, besides less technical exposition, articles on the teaching of mathematics and administrative issues. This informality is reflected by the choice of cover picture for the edition containing Weinstein's article. Next to the title of this article one sees a photograph of a herd of zebra. No explicit explanation is offered for its presence, nor is one needed. The received account as to why zebras sport stripes is that when they stand in a herd, a charging lioness is presented with a strongly patterned visual array, making it very difficult for her to detect the outline of a single member of the herd. The rationale for the choice of this picture, in which one imagines Weinstein played a part, rests in his idea that groupoids are better than groups at detecting the inner symmetry of patterns of this kind. This idea Weinstein explicitly illustrates in the article itself with a discussion of the symmetries of a set of bathroom tiles. In contrast to this rather mundane concern of the mathematician contemplating the pattern of the grouting while enjoying a soak, the cover picture makes clear that such inner symmetry is a matter of life and death. As any zebra will tell you, 'symmetry capturable by groupoids but not by groups saves lives'.

Let us now consider the definition of a groupoid and its motivation.[7] A groupoid is composed of two sets, A and B, two functions, a and b, from B to A, and an associative partial composition, $s{\cdot}t$, of pairs of elements of B with $a(s) = b(t)$, such that $a(s{\cdot}t) = a(t)$ and $b(s{\cdot}t) = b(s)$. Furthermore, there is a function, c, from A to B such that $a(c(x)) = x = b(c(x))$ and such that $c(x){\cdot}s = s$ for all s with $b(s) = x$ and $t{\cdot}c(x) = t$ for all t with $a(t) = x$. Finally, there is a function, i, from B to B such that, for all s, $i(s){\cdot}s = c(a(s))$ and $s{\cdot}i(s) = c(b(s))$.

This may seem like a highly convoluted definition, but it can be illustrated simply in Brown's picture. We simply take A to be the set of cities, while B is the set of trips. The start and finish of a trip are given by applying a and b, respectively. Applying c to a city results in the staying-put trip. Finally, i sends a trip to the same trip in reverse.[8] This illustration should prompt anyone acquainted with category theory to realise that a groupoid may be defined concisely in its terms. Indeed, a groupoid is just a small category in which every arrow is invertible. This much curter definition points to an important association of groupoid theory with category theory, as we shall see later. From this perspective, groups can be seen to be special cases of groupoids, that is, they are groupoids with only one object. Alternatively, in terms of the definition above, a group may be represented as a groupoid in which the set A is a singleton, and where B corresponds to the set of group elements seen as permutation maps on the group.

9.3 HOW GROUPOIDS COMPARE WITH GROUPS

The fact that groups are just a type of groupoid raises the possibility that groupoids comprise a more conceptually basic variety of object. The first mathematician into whose consciousness groupoids explicitly appeared seems to have been H. Brandt. In his research on quaternary quadratic forms he found that he could define a composition on classes of forms, but unlike in the binary case where a group is involved, this composition was only

---

[7] Note that the term 'groupoid' is also used to denote a set with a binary operation on it satisfying no further conditions. This minimalistic structure is used by Saunders Mac Lane as an example of what he calls a 'mathematical dead end' (Mac Lane 1992: 10).

[8] Again these trips are being considered here only 'up to homotopy'. Higher-dimensional groupoids can used to maintain the distinction between homotopic paths. Note also that one-way streets are being overlooked.

partial. He named the corresponding structure a *gruppoid* (Brandt 1926). As this was a continuation of a programme begun by Gauss, quite possibly groupoids might have been defined earlier. One can speculate that, had the course of history run differently, we would find what we now call groups being designated by some epithet as a type of what we now call groupoid, rather than having, as is now the case, groupoids seen as 'not-quite-groups'. On the other hand, it seems very likely that category theory would have had to have been invented first.

Historical counterfactuals do not take us far. What we need now is a comparison of the characters of the group and groupoid concepts. Given that the group structure *had* already been isolated, our question is how much was there to gain by generalising to groupoids. Groupoids will have to confront the charge that anything they achieve was already inherent in the idea of groups. Before they are allocated some of the goodwill earned by their relatives, they will need to prove sufficiently different to enable their users to do new things and to do old things more straightforwardly. To the extent that the monoid concept is a generalisation of the group concept, in that the requirement that each element has an inverse is dropped, one might imagine that it might be in a similar situation. However, it is easy to argue that the character of monoids is very different and that monoids will have to make their own way in the world. In that the idea of an inverse is central to any concept of symmetry, groupoids, as their name was designed to suggest, would appear to be lesser distortions of groups than are monoids.

In favour of the idea that groupoids are similar in spirit to groups, we find that only a small modification is required:

Thus the groupoid I, which at first sight seems unworthy of notice, plays a key role in the theory of groupoids, and in applications. A failure to extend group theory so as to include the use of I, on the grounds that I is a trivial object of only formal interest, is analogous to failing to use the number 0 in arithmetic, a failure which in fact held back mathematics for centuries. Of course, if you allow I, then in effect you allow all groupoids since any groupoid is a colimit of a diagram of copies of I, in the same way as any group is a colimit of a diagram of copies of $\mathbf{Z}$. (Brown 1987: 121)

The groupoid I is composed of two objects, identity arrows and an arrow passing in each direction between the objects. Think of two cities with a single road between them. Thus, overcoming our resistance to groupoids is likened to that monumental moment when zero was recognised as a number – a small change with large ramifications.

The opposition[9] can pick up on the size of this change, however. One of the few explicit criticisms of groupoids portrays them as only a minor variant of groups, the real essence of the notion of symmetry. After all, it is the case that a (transitive) groupoid is isomorphic to the product of the vertex group at any object and the coarse groupoid on the elements of A, that is, the groupoid with a single arrow between each pair of elements of A. To carry out this reduction in terms of Brown's illustration, for each city select a path leading to it from Bangor. Then any trip from, say, London to Birmingham can be recreated from the designated paths from Bangor to each of these two cities and the group of round trips from Bangor. The complexity of a groupoid appears to be already contained within any of its isomorphic vertex groups, the arrows looping around at a given point. Compare this to the intricate structure theorems of groups themselves or of von Neumann algebras.

A related point is that the naming of examples of a certain class of entity acts to give the definition of that class a greater sense of importance. For instance, the largest of the sporadic simple finite groups is known as the *monster*. In addition to its vast size it has recently received additional fame through the connections established between its representations and the j-function as we saw in chapter 4. Elsewhere, we can find noted von Neumann algebras such as the hyperfinite type II_1 factor and among Lie algebras E_8 attracts much interest. With groupoids, on the other hand, no individual stands out that is not a group. One might point to the simplest groupoid which is not a group, the one we denoted I above, but it does have a very simple structure.

Against the 'trivial classification' criticism, Brown and Weinstein produce the same two counter-arguments, the first of which runs to the effect that if this criterion is to be applied rigorously then important entities such as finite vector spaces become vacuous as they are categorised simply by a natural number. What is vital for vector spaces is the linear maps between them. So it is with groupoids. Notice also that no finite vector space stands out.

The second line of defence argues that especially interesting things happen when you add extra geometric structure to groupoids. Groupoids come in several varieties: topological, measurable, differentiable, Lie, Poisson, symplectic, quantum, algebraic, etc. The geometric structure often interacts with the groupoid structure in a more complicated way than in the corresponding situation with only a group structure.

[9] Most of the opposition takes the form of a reluctance to use groupoids. Some comes in the form of anonymous referees' reports on grant proposals. We shall see some of the small amount of explicit opposition below.

I have yet to read or hear of any riposte to these lines of defence. This of course does not mean that such a thing is impossible, but it does indicate that mathematicians, unlike philosophers, have no particular inclination to engage in sustained argumentative activity. I am inclined to believe that this is due to a deficiency in mathematical training, rather than because it is unnecessary. Lakatos drew a similar conclusion when through the voice of the student Gamma he wonders:

Why not have mathematical critics just as you have literary critics, to develop mathematical taste by public criticism? (Lakatos 1976: 98)[10]

However, while it is difficult to discover arguments passing through several turns of criticism and defence, it is quite straightforward to find a considerable range of lines of argument put forward to support a construction.

Let us now turn to consider some of the reported advantages of groupoids. There is only space in this chapter to touch on a few of these advantages, which may be classed as follows:

(1) As generalisations of groups, they fully capture the one-dimensional aspects of a situation

(2) As generalisations of equivalence relations, they cope well with the symmetries of 'bad' inhomogeneous spaces

(3) Applications in physics for groupoids have been found which go beyond the use of group theory

(4) Higher-dimensional groupoid theory is richer than higher-dimensional group theory and allows new geometric features to be measured.

9.4 THE FULL EXPLOITATION OF ONE-DIMENSIONALITY

Groups do not fully exploit all the path-like behaviour that is present in a situation, because they do not capture the intermediate stages of reversible processes. We can see this in the following example. Recall from algebraic topology that the fundamental group of a space at a base point is the set of classes of closed paths in the space beginning (and ending) at that point, where two paths belong to the same class if one can be continuously deformed to the other within the space.[11] Van Kampen's theorem tells you that

---

[10] See also Brown (1994: 50): 'Does our education of mathematicians train them in the development of faculties of value, judgement, and scholarship? I believe we need more in this respect, so as to give people a sound base and mode of criticism for discussion and debate on the development of ideas.'

[11] It is not essential to gain a thorough understanding of the mathematics which follows. For those who wish to see a more leisurely presentation of this material I can recommend Gilbert and Porter (1994).

if you know the fundamental groups of two spaces, U and V, and of their intersection, $U \cap V$, at x, a base point in the intersection, and if all three spaces are path connected, then you can calculate the fundamental group of $X = U \cup V$ at x. This theorem is used to pass from the fundamental group of simple spaces such as the disc and circle to more complicated spaces. The fundamental group of the disc for any base point is trivial, since any loop can be deformed to the constant loop. In the case of the unit circle, loops which pass a given number of times around the origin are equivalent, so may be classified by an integer. Composition of paths then corresponds to integer addition.

Van Kampen's theorem tells you that the fundamental group of the union of U and V has as generators those of U and V, but that in addition to the relations already in place new ones may be imposed. These arise from equating the two representations of a loop situated in the intersection, according to whether it is viewed as belonging to U or to V. For example, a torus may be taken as the union of two spaces: an open rectangle, U, and a union of two annuli, V. The fundamental group of U is trivial and that of V is the free group on two generators, since it is retractable to the join of two circles. The loop c around the intersection of U and V is collapsible in U, but is homotopic to the path $a^{-1}b^{-1}ab$ in V. Therefore, in the fundamental group of the union the latter path must be put equal to the identity, or in other words, the relation $ab = ba$ is imposed and we can conclude that the fundamental group of the torus is the Abelian group on two generators:

$$\pi(U, x) = \; <|>, \pi(V, x) = \; <a, b\,|> \text{ and } \pi(U \cap V, x) = \; <c\,|>.$$

In U, $c \sim$ constant path at x. While in V, $c \sim a^{-1}b^{-1}ab$.

Therefore, $\pi(\text{Torus}, x) = \; <a, b\,|\, a^{-1}b^{-1}ab> \; = Z \oplus Z$.

But how can the result that the fundamental group of the circle is isomorphic to the integers under addition be derived? There are several ways of doing this, none of which is as straightforward as might be expected for such a basic shape. It might have been hoped that it could be found by applying van Kampen's theorem to the circle seen as the union of two overlapping open intervals:

That this will not be possible is clear from the observation that, in the words of Rogers and Hammerstein, 'nothing comes from nothing' – the fundamental groups of U and V being trivial and so providing no generators. The problem arises from the fact that the intersection of U and V is not connected, but is composed of two disjoint intervals. An arbitrary choice must be made as to which component will contain the base point. As we saw earlier with the disconnected United Kingdom, fundamental groups do not cope well with such spaces as they can measure only one connected component. This presents something of an anomaly since open regions of n-dimensional space are the basic building blocks of manifolds in modern topology and geometry. There is, of course, nothing wrong with the various ways of establishing the fundamental group of the circle. No Lakatosian proof analysis conducted on any of them will discover a counter-example. One also cannot deny the interesting connections with many other branches of mathematics, e.g., the winding number of a function about a point in complex analysis. The point is, however, that the van Kampen philosophy just ought to work there. To call this case a 'heuristic counter-example' would be to stretch the meaning of the term beyond that given it by Lakatos (1976: 83), but I think he would have approved.

In fact it can be made to work but only if one extends the fundamental group idea to allow loops at several base points and paths between them. One for each component would be enough. But after this extension we shall no longer be dealing with a group since composition will not be possible for each pair of paths. The fundamental groupoid of a space X with respect to a given set of base-points has this set as A and equivalence classes of paths between two such points as the elements of B. Groupoid status is assured owing to the fact that each path may be run backwards.

One can now prove a van Kampen theorem for fundamental groupoids. I shall not enter into details, but note that its phrasing in category theoretic terms is very simple. The category of topological spaces has *pushouts*. A pushout may be thought of as a kind of sum of two objects which identifies or keeps separate precisely what ought to be identified or kept separate. In particular, the pushout of the injections of the intersection, $U \cap V$, into U and into V is their union, $U \cup V = X$. The fundamental groupoid construction provides a *functor* from the category of topological spaces to the category of groupoids which preserves pushouts. With U and V overlapping intervals forming a circle, and x and y points in the components of the intersection, we have:

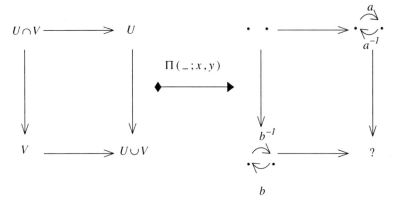

This tells us that insofar as we are interested in one-dimensional data, much about the compositional structure of topological spaces has been captured algebraically. Indeed, the fundamental groupoid of the circle with a pair of base-points is the pushout of the diagram of groupoids (denoted by the question mark) and one can prove that the vertex group at each object in this groupoid is isomorphic to the infinite cyclic group.[12]

Although groups are more familiar to mathematicians, the restriction to one base point may also lead to an unwieldy presentation in terms of generators and relations, rather as it would be inconvenient to view all British road trips in relation to Bangor. As Grothendieck remarks on the benefits of groupoid presentations:

> people are accustomed to work with fundamental groups and generators and relations for these and stick to it, even in contexts when this is wholly inadequate, namely when you get a clear description by generators and relations only when working simultaneously with a bunch of base-points chosen with care – or equivalently working in the algebraic context of *groupoids*, rather than groups. Choosing paths for connecting the basepoints natural to the situation to one among them, and reducing the groupoid to a single group, will then hopelessly destroy the structure and inner symmetries of the situation, and result in a mess of generators and relations no one dares to write down, because everyone feels they won't be of any use whatever, and just confuse the picture rather than clarifying it. (Quoted in Brown 1987: 118)

Groupoids even provide new information about groups themselves, because they possess some important properties not shared by groups. For example,

---

[12] The identity arrows of the groupoids have not been shown. The composition of arrows *a* and *b* is an arrow from *x* to itself. As nothing tells us to equate either it or any iterate of it to the identity arrow at *x*, we do not. Indeed, it forms a generator for the vertex group at *x*. Notice how the simple groupoid *I*, which marks the gap between groups and groupoids, crops up here.

the category of groups cannot support several useful constructions:

> One of my hopes in preparing the text was to convince students of group theory that it is often profitable to cross the boundary between groups and groupoids. The main advantage of the transition is that the category of groupoids provides a good model for certain aspects of homotopy theory. In it there are algebraic analogues of such notions as path, homotopy, deformation, covering and fibration. Most of these become vacuous when restricted to the category of groups, although they are clearly relevant to group-theoretical problems. (Higgins 1971: vii)

A further significant flaw with groups is that:

> One of the irritations of group theory is that the set Hom(H, K) of homomorphisms between groups H, K does not have a natural group structure. However, homotopies between homomorphisms of groupoids H, K may be composed to give a groupoid HOM(H, K) with object set Hom(H, K). (Brown 1987: 122)

This construction leads to a groupoid isomorphism HOM($G \times H$, K) HOM(G, HOM(H, K)), an example of a very widespread structural law which is found even in the simplest theories, such as arithmetic, $a^{(b \times c)} = (a^c)^b$, and propositional logic, A & $B \vdash C$ if and only if $A \vdash B \rightarrow C$. The desirability of this property is also the reason some topologists give for working with compactly generated topological spaces.[13] Use of this construction again provides information about groups.

9.5 GROUPOID ALGEBRAS USED TO COMPENSATE FOR BAD SPACES

First let us consider a simple way in which the algebra of complex $n \times n$ matrices may be reinterpreted from a groupoid perspective. Take the pair (or coarse) groupoid $A \times A$, where $A = \{1, 2, 3, \ldots, n\}$. This is the groupoid where for each pair of members of A there is a single arrow passing from the first to the second. Next, take the algebra of complex valued functions on the arrows of this groupoid. A type of multiplication known as convolution may now be defined on these functions, generalising a similar construction used for groups. The value of the convolution of two such functions, f and g, denoted $f * g$, on a pair (i, k) is the sum over j of products of the form $f(i, j) \cdot g(j, k)$. This is completely equivalent to matrix multiplication, where if the ijth entry of a matrix, M, is $f(i, j)$ and similarly for a matrix N and the function g, then the matrix corresponding to $f * g$ is simply $M \cdot N$.

---

[13] Cf. Mac Lane (1971: 184).

Convolution algebras can be defined similarly for all groupoids. In view of the importance of group convolution algebras, which over the integers underlie the harmonic analysis of Fourier series, we can see the potential for this generalisation. Indeed, groupoid convolution algebras play a major part in the field of non-commutative geometry. This field is based on the observation that a commutative algebra may be construed as the collection of functions on a space. By analogy the non-commutative algebras are seen as arising from functions on a *non-commutative* space. Such a space is often characterisable as the orbit space of a groupoid. To recapture a space of functions on what might be termed a 'bad' space with an inadequate collection of ordinary set-based functions, the focus is shifted from the space to the groupoid representing it, and from there to a suitable convolution algebra.

To take a simple example, consider the topological space formed from two intervals, $\{0\} \times [0, 1]$ and $\{1\} \times [0, 1]$, by identifying pairs of points $\{(0, a), (1, a)\}$ for all $a \neq \frac{1}{2}$. This looks very much like just one interval, except for a small split half-way along it. It is a perfectly legitimate topological space, although not Hausdorff. But then if we try to characterise the space by the continuous complex functions it can support, we find that the points $(0, \frac{1}{2})$ and $(1, \frac{1}{2})$ cannot be distinguished. From the point of view of continuous functions the space is indistinguishable from a simple interval. However, this space can be reformulated as the orbit space of a topological groupoid where the objects are the points along the two intervals, and besides the identity arrows, there are pairs of inverse arrows between the points to be identified. This groupoid is not equivalent (homotopic) to the groupoid of identity arrows on an interval, as becomes apparent by the difference between their convolution algebras of complex continuous functions. This may be represented in the case of the 'bad' non-Hausdorff space as the algebra of 2×2 matrices of continuous complex functions on $[0, 1]$ which are diagonal at $\frac{1}{2}$.

This is a simple example of an important use for groupoids, which Weinstein claims:

leads us to the following guiding *principle* of Grothendieck, Mackey, Connes, Deligne, . . .
Almost every interesting equivalence relation on a space B arises in a natural way as the orbit equivalence relation of some groupoid G over B. Instead of dealing directly with the orbit space B/G as an object in the category S_{map} of sets and mappings, one should consider instead the groupoid G itself as an object in the category G_{htp} of groupoids and homotopy classes of morphisms. (Weinstein, 1996: 748, my emphasis)

Here we see once more the expression of a broad principle, which is what I took earlier to be the mark of the conceptual. Notice the vagueness of the wording. There is plenty of scope here for argument about what counts as 'almost every' and as 'interesting' and as to whether it really is so very 'natural'. As we saw in chapter 8, this vagueness is reminiscent of the kind of language with which the aims and means found at the heart of a research programme are articulated.

Groupoids as generalisations of sets, equivalence relations, groups and group actions permit a unified reformulation of these concepts. This fact by itself is not sufficient reason to adopt them; concepts must work harder to pay their way. Whether a reformulation constitutes a clarification and whether a unification may be considered important are two deep questions. For Weinstein the approach to orbit spaces outlined in this section *has* led to an important unification and he claims that Alain Connes's book on non-commutative geometry, which utilises this construction:

shows the extent to which groupoids provide a framework for a unified study of operator algebras, foliations, and index theory. (Weinstein 1996: 745)

9.6 APPLICATIONS AND OLD CONJECTURES

Having applications in the sciences, computing or engineering is incontrovertibly a good thing for a piece of mathematical theory. However, questions remain concerning the importance of the application and whether a particular theory is indispensable in a given application. In the case of groupoids, we hear in an announcement for a 1998 conference, 'Groupoids in Physics, Analysis and Geometry', that:

The uses of groupoids in physics come from two main sources. The first is Alain Connes' theory of noncommutative geometry, in which groupoids are a main source of examples of noncommutative spaces. This theory is being studied very actively by physicists, and by mathematicians. Bellisard's work studying the quantum Hall effect via noncommutative geometry has led to the study of connections between solid state physics and noncommutative geometry models associated with tilings.

The second major source of the use of groupoids in physics is the general theory of quantization in mathematical physics. A theory of quantization has been introduced by V. Maslov and A. Karasev, and a version due to Alan Weinstein has been actively developed by him and his collaborators. One step in this program is to associate a symplectic groupoid to a given Poisson manifold. (Kaminker 1998)

As for the first of these uses, Bellisard, a solid state physicist, uses non-commutative geometry to explore the non-commutative Brillouin zone of an aperiodic medium. While discussing the way mathematicians succeeded in capturing some forgotten intuitions of Heisenberg in the context of the C^*-algebra approach to quantum mechanics, he claims that:

The breakthrough went with the notion of a groupoid . . . which is nothing but the abstract generalization of the notion of transition between stationary states as defined by Bohr and Heisenberg. (Bellissard 1992: 551)

This relates to the reformulation we saw above of matrix algebras as the convolution algebras of groupoids. It is Connes's position that for Heisenberg the groupoid idea came first, albeit implicitly, and hence his requirement of matrices. As he explains in a section of his book (1994: 33–9) written to 'remove this prejudice [towards groupoids]' (1994: 7), the groupoid idea is present in the case of an electron's transitions between energy levels in the atom – the transitions from level i to level j and from level k to level l may be composed iff $j = k$.

With the physical world providing only a very indirect constraint on mathematical theorising, mathematicians have worried that pieces of research, although perfectly correct, may be of little or no value. They have, therefore, sought ways internal to mathematics of adjudicating whether a theory is on course. One way of doing this is to set up, as Hilbert famously once did, a series of problems to be solved. Then a theory's success in solving any of these problems can be taken as a token of its worth. Thus:

Often a test for the value of a new theory is whether it can solve old problems. *De facto*, this limits the freedom of a mathematician, in a way which is comparable to the constraints imposed on a physicist, who after all doesn't choose at random the phenomena for which he wants to construct a theory or devise experiments. (Borel 1983: 14)

One senses here the concern that without such constraints mathematicians may find themselves wandering aimlessly through a world of mathematical possibility.

Now, on this score there appear to be no clear successes for groupoids,[14] but remember that groupoids can be used to discover new properties about familiar things, namely, groups. Just because these properties were unforeseen, and so no conjectures made about them, seems to be no reason to mark groupoids down.

---

[14] It may be argued, however, as Brown has, that Grothendieck's reliance on groupoids means that they are due some credit for Wiles's proof of Fermat's Last Theorem.

As a second response, we might say that while it is reassuring that a theory solves famous old problems, mathematics must also be about opening up new areas by the elaboration of mathematical ideas. Frequently, for those venturing into unknown territory there is no shortage of constraints resembling the 'old problems' one. For example, you may find that a definition that seems to go in the right direction unexpectedly makes contact with older work, or that the method you are using to overcome obstacles which are preventing you from performing a construction analogous to an earlier one gives you a much clearer picture of the whole domain. We now turn our attention to some new areas.

9.7 NEW PROSPECTS: HIGHER-DIMENSIONAL ALGEBRA

The fundamental group of a space need not be Abelian. Imagine yourself based at the crossover point of a figure of eight. The path which takes you clockwise round the upper loop, then around the lower one is not equivalent to the path taking the loops in the opposite order. However, higher homotopy groups *are* always Abelian. Here, rather than throwing loops into our space to see what we can catch, we are throwing spheres (two-dimensional and higher). Just as we can see a loop belonging to the fundamental group as a line where the endpoints are identified, we can see an element of the second homotopy group as a map of a square into the space where the perimeter gets mapped to the base point. Think of a net having being cast by a fisherman, who now holds its opening. Then we can set up a composition in two directions, corresponding to the two dimensions of a square.

Let us give an idea of what happens when we compose in one direction. Imagine two square nets joined along one edge and pinned to the table along their perimeters. Push all the raised part of the left net into its upper half so that the rest lies flat, and all the raised part of the right net into its lower half. Then make the raised parts swap sides and permits them to reform their original shapes. You may be able to see that the two multiplications coincide in a single commutative operation.

The question then arises as to whether this commutativity is due to the higher dimensional homotopic nature of spaces or whether it is a failure on the part of groups to capture this nature. On the face of it there is no reason to expect homotopy to become simpler in higher dimensions, suggesting that the fault lies with the algebra, which must be refined to detect deeper features of geometric reality. As part of the process of capturing these deeper features, in the 1940s the topologist J. H. C. Whitehead devised what are known as crossed modules. Brown has succeeded in using them in this way and he notes that:

information about even such an apparently simple computation as a second absolute homotopy group of this mapping cone is tightly bound to information on crossed modules. There is at present no alternative description [to crossed modules] of this second homotopy group in algebraic terms. This highlights some basic difficulties of homotopy theory, and also suggests that homotopy theory is an essentially non abelian subject. The abelian homotopy groups, even as modules over the fundamental group, give only a pale shadow of the homotopical structures. (Brown 1999: 32)

Now, crossed modules turn out to be equivalent to groupoid objects within the category of groups, i.e., groups on which there is a compatible groupoid structure. Unlike in the purely group theoretic case, the two structures interact non-commutatively.

The next step is to look for double groupoids, groupoid objects within the category of groupoids – or, if you prefer, two interacting groupoid structures. The simplest way to catch a glimpse of what is happening here is to think of mapping a square into a topological space as we did above, but this time with no restrictions on where its perimeter lands. We still have multiplications running in two directions but, in the spirit of groupoids, only if the paths corresponding to the adjoining sides of two square are equal. Brown managed, after years of effort, to achieve a van Kampen-style theorem in two dimensions. These mark some early steps of an enormous programme we shall discuss in chapter 10.

We should note that multiple and higher-dimensional groupoid theory has not penetrated into the non-commutative geometry mentioned in sections 9.5 and 9.6, although it is starting to be used in differential geometry (see, e.g., Mackenzie 1992). In view of the fact that Charles Ehresmann was exploring such ideas in the late 1950s, we may wonder why the development has been so slow. Has there been an undervaluing of the conceptual?

9.8 THE CONCEPTUAL AND THE NATURAL

Recall my quoting earlier Brown making 'the heretical suggestion that the *natural concept* is that of groupoid rather than group' (Brown 1999: 4, my emphasis). The philosophical treatment of the notion of a mathematical concept is still to be done, but it is interesting to note that the category theorist William Lawvere, co-author of *Conceptual Mathematics* (1997), has expressed the view (Brown 1987: 129) that the term *group* should be taken to refer to what is now covered by *groupoid*. Most mathematicians will find this hard to accept, having been taught to accept the group concept as the natural one. Let us approach these matters by discussing the idea of *naturalness*.

The epithet 'natural' is never far from mathematicians' lips when they describe their favourite constructions. It even appears in mathematical terms such as natural number, natural transformation and natural deduction. Only for the first of these can some connection with the physical world be claimed, as the natural numbers constitute possible responses to questions of the kind 'How many elephants are there in this National Park?'. Mathematicians sometimes play humorously on this idea. In what must be one of the wittiest mathematics textbooks ever written, Frank Adams discusses the situation where two spaces have the same homology groups, yet their fundamental groups are 'wildly' different. He tells us that one of the spaces can even be taken to have trivial higher homotopy groups:

By now we have theorems saying that this situation is common; Kan and Thurston show that given almost any space Y, you can approximate it homologically by an Eilenberg-Mac Lane space $EM(\pi, 1)$ for some weird and artificial group π. However, we should perhaps be more concerned with cases where this situation arises in *nature*. (Adams 1978: 84, my emphasis)

Without wishing to labour the point, you are not going to meet with this situation while on safari in a National Park. You won't even meet with it while doing theoretical physics. But you may encounter it, without artificially engineering it, while working in reasonably well frequented regions of mathematics. Adams's sentiment is that for a type of construction to be worth defining or for a type of situation to be worth describing, there ought to be examples readily available.[15]

For Adams, if an instance of the situation he describes occurs and he decides not to count it as arising in nature, he is not thereby banishing that instance from its fellows. This may be contrasted with what Lakatos (1976: 23) designated as 'monster-barring', when a proposed counter-example which may be thought to have refuted a claim about a class of entities is declared not to belong to that class. Here, the 'monster' is unnatural – it does not have in its nature what it takes to be a member. Adams's reaction is more

[15] A similar idea is expressed by Robert Solomon when he points out that, despite the fact that the majority of finite groups are nilpotent of nilpotence class 2, and so far from being simple, 'experience shows that most of the finite groups which occur "in nature" – in the broad sense not simply of chemistry and physics, but of number theory, topology, combinatorics, etc. – are "close" either to simple groups or to groups such as dihedral groups, Heisenberg groups, etc. which arise naturally in the study of simple groups' (Solomon 2001: 347).

typical of the contemporary mathematician, who knows enough about the conceptual twists and turns that have occurred in the discipline since the mid-nineteenth century not to take talk of unnatural monsters too seriously and illustrates an important point about the changing conceptions of mathematicians towards the role of definitions. It would be a valuable exercise to make comparisons with claims of naturalness from earlier times.

We can see the modern attitude illustrated in the following example from non-commutative geometry. While discussing the set, X, of Penrose tilings,[16] Alain Connes notes that, although it is clearly a spatial entity, when it is treated with the classical tools of point-set topology it cannot be distinguished from a point. This is because among the peculiar properties of X we find that, given any two distinct tilings, a finite portion of one of them of whatever size will be found occurring infinitely often within the other. Hence:

[t]he natural first reaction to such a space X is to dismiss it as pathological. (1994: 6)[17]

This may sound rather like monster-barring, but what Connes means here is that:

To a conservative mathematician this example might appear as rather special, and one could be tempted to stay away from such spaces by dealing exclusively with more central parts of mathematics. (1994: 94)

So, it is not a question of excluding X by modifying the definition of a topological space. Either one accepts it as an odd sort of space and then ignores it or, like Connes, one brings new tools to bear upon it, in this case a convolution algebra on the associated groupoid. The situation may be summarised well by describing X as a *heuristic counter-example* to the notion that classical topology is adequate to deal with all topological spaces.

Another way of arguing for the naturalness of a concept is in terms of the inevitability of its discovery. There seems to be a widespread feeling that

---

[16] These are the quasi-periodic tilings of the plane with local 5-fold symmetry whose patterns have been found to occur in the natural world in what are termed *quasi-crystals*. X may be interpreted as the orbit space of a groupoid.

[17] Notice here how 'natural' is being used about the mathematician rather than about the mathematical entity. Perhaps *natural deduction* involves both. It turns the reasoning processes of mathematicians into an entity which may be investigated mathematically.

however mathematics was to reach anywhere near the level of sophistication we see today, a basic concept such as that of a group was bound to be formulated, while there is disagreement over whether the same could be said for groupoids. Powerful evidence that a concept was inevitably going to be forged is to show that it was required independently by researchers working in different fields. A convincing case can be made that this was so for groups. As for groupoids, after their introduction by Brandt in 1926, researchers have for their own reasons deemed it worthwhile to introduce them into the theory of field extensions, non-commutative ring theory, algebraic logic,[18] partial differential equations, category theory, differential geometry, differential topology, foliations, non-Abelian cohomology and ergodic theory.

Independence of use is most marked when the researcher coins a new name for the concept as when George Mackey working in ergodic theory used the term 'virtual group' to refer to what amounts to a groupoid. A more vivid illustration of this phenomenon is reported in a survey article on Lie algebroids and Lie pseudoalgebras[19] by Kirill Mackenzie. He remarks (1995: 100) that the notion of a Lie pseudoalgebra had been devised independently by well over a dozen researchers in almost identical fashion, each with a different name.

The mathematicians' notion that some ideas are fundamental and will inevitably emerge reveals a degree of faith resembling that motivating the scientists' discursive line, treated by Gilbert and Mulkay (1984), that 'Truth will out'. These ideas are deemed to possess such intrinsic value that they can overcome the vagaries of the human research effort. The strongest form of this sentiment would maintain that some concepts will necessarily appear and that, by virtue of their nature, they rather than the user will determine their use. Opposed to this faith that methods of research will not stand in the way of important ideas and their proper deployment is the notion that even good ideas that have at some time surfaced into the awareness of a mathematician may be lost to future generations. For Gian-Carlo Rota this is no rare event:

On leafing through the collected papers of great mathematicians, one notices how few of their ideas have received adequate attention. It is like entering a hothouse and being struck by a species of flowers whose existence we did not even suspect. (Kac *et al.* 1986: 1)

---

[18] In the early 1950s, Jónsson and Tarski required *generalised Brandt groupoids* to capture the calculus of binary relations. These were not required to satisfy transitivity.

[19] The former are related to Lie groupoids as Lie algebras are to Lie groups and are special cases of the latter.

Presumably, then, for Rota some of these powerful ideas may be lost for a long time, and possibly forever.

This picture offers the mathematician the opportunity of presenting their work as allowing the recovery of some of this lost treasure. Some mathematicians are appealed to more than others in this respect. Someone like Sophus Lie working in a branch of geometry at a time when standards of rigour had not become well established, but when mathematicians were closer to 'nature', makes for an excellent target. Hence:

[t]he concept of groupoid is one of the means by which the twentieth century reclaims the original domain of application of the group concept. The modern, rigorous concept of group is far too restrictive for the range of geometrical application envisaged in the work of Lie. (Mackenzie 1987: vii)[20]

There are various ways of responding to claims of naturalness. I may suggest a new concept to be the natural development of an earlier one, or the natural idea on which to base an attack on an important problem, or the natural way to illuminate some phenomenon in the physical sciences. You reply by claiming it to be an unnecessary modification of a perfectly serviceable idea, with little to be gained from its acceptance other than as a boost to my publication record. A colleague then chips in with her view that while she does not believe it to have achieved what I claim, it may prove useful as a temporary measure, and may lead to a better reconceptualisation of the field. This threefold distinction – fundamental, convenient and pointless – is quite common. In the following quotation we see groupoids consigned to the middle category:

[definitions] like that of a group, or a topological space, have a fundamental importance for the whole of mathematics that can hardly be exaggerated. Others are more in the nature of convenient, and often highly specialised, labels which serve principally to pigeonhole ideas. As far as this book is concerned, the notions of category and groupoid belong in this latter class. It is an interesting curiosity that they provide a convenient systematisation of the ideas involved in developing the fundamental group. (Crowell and Fox 1977: 153)

Naturally, Brown sees this as much more that an 'interesting curiosity'. Rather, the elegance of the systematisation was read by him as a clue that groupoids could play a very large role in this area.

The *convenient* class is very broad. Crowell and Fox judge groupoids to be at the lower end, bordering on the pointless. Meanwhile, in a discussion of

---

[20] Lie is also selected by two exponents of synthetic differential geometry who claim to be able to allow his intuitive reasoning to be fully captured in a rigorous framework (Moerdijk and Reyes 1991).

the role of groupoids in differential geometry, Kumpera has them straddling the boundary between the fundamental and the convenient:

Bundles are of course extremely useful objects but, as Ehresmann would probably say, groupoids are somehow closer to the truth. As for connections, they are an extremely useful algorithm whereas groupoids (and algebroids) are an extremely useful concept. (Kumpera 1988: 359)

Rota goes as far as to say of the notion of groupoids that it 'is one of the key ideas of contemporary mathematics' (Bergeron *et al.* 1997: vii).

We see here the idea of a split between something intrinsically worth studying and something valuable as a tool to be used in the study of something else, even if this means that the tool needs to be studied to know if it is up to the job.[21] But this distinction is not permanent. In the course of time the status of mathematical entities may change from being viewed as useful tools to becoming fully fledged objects. In the search for a solution to a problem, means are introduced which can then become interesting in their own right, and further means will then be necessary to study them in turn:

Once we have a genuine need for some mathematical idea as a matter of language, that idea has arrived; it is hardly necessary to discuss its status as a useful technical tool. (That sentence is not intended to exclude the possibility that some authors may try to introduce language we can do without.) (Adams 1978: 79)

The criticism which may meet this kind of promotion is that means have been unjustifiably raised to the level of ends.

Category theory is often picked out for this treatment as when Miles Reid remarks:

The study of category theory for its own sake (surely one of the most sterile of all intellectual pursuits) also dates from this time; Grothendieck himself can't necessarily be blamed for this, since his own use of categories was very successful in solving problems. (Reid 1988: 116)[22]

As a contemporary algebraic geometer, Reid could not possibly deny category theory its 'useful' status. Even in a textbook aimed at undergraduates (Reid 1988) he allows it to make an occasional appearance. What he appears

---

[21] One might be tempted to equate the class of that which is worthy of study for its own sake with the *natural*, but mathematicians might easily talk of a piece of what they term *machinery* as natural. Topologists actually refer to the apparatus for converting between spectra and spaces as 'machinery'. See Adams (1978, ch. 2). I have sometimes been asked why the methodologist need bother studying (difficult) contemporary theory development. The answer is clear: if she wants to find out how today's mathematical 'technology' works, she has no choice.

[22] For robust rebuttals of Reid's remarks see Brown (1994: 49).

to be criticising here is the view that categories are like meta-Abelian groups, braids, projective varieties or sets, entities to which a mathematician today may devote her whole career without being required to demonstrate their applicational virtues. His is probably a majority position at the present time,[23] but again within this position finer distinctions may be made ranging from those who think category theory works well as a convenient means of representing a body of theory to those who think its principles can at times provide very strong guidance for future research. Moreover, promotion for categories may be imminent. With the race to develop a theory of weak *n*-categories currently proceeding at a frenetic pace, one may argue that important entities of intrinsic conceptual interest are being carved out.

This mention of category theory leads us to a narrower, more technical, sense of the term 'natural'. As we saw above, the classification of a transitive groupoid requires the choice of an object, and the choice of an arrow from that object to each other object. These choices cannot, however, be said to be natural, insofar as there was no good reason to favour them over any others. In other words, the reduction of a groupoid to a group suffers from the need to make an arbitrary, or unnatural, choice. This might remind the reader of a similar unforced choice in the theory of vector spaces. To establish an isomorphism between a finite vector space and its dual, the space of linear maps to the ground field, one must make an arbitrary choice of a basis. By contrast, when it comes to establishing an isomorphism between a vector space and its double dual, there is a natural map, namely, the map which sends a vector to the map equivalent to evaluation at that vector. This last example is part of the category theoretic folklore. It provided Eilenberg and Mac Lane with the notion of a natural transformation between functors, in this case between the identity and double dual functors on the category of vector spaces.

Here we see again the idea that it is preferable not to make arbitrary choices. It accords with the notion that one should not privilege a member of a collection without good reason. Recall that we constructed the fundamental groupoid of a space precisely to avoid creating a privileged base-point. This kind of privileging occurs in the notion of a principal fibre bundle from differential geometry where one fibre is singled out. A more even-handed or 'democratic' treatment of this very important geometric idea is to work with Lie groupoids, those for which the sets we have

---

[23] Of course, the positing of a piece of research as 'for its own sake' is open to challenge.

denoted *A* and *B* are smooth manifolds satisfying some further conditions. This is not just done for convenience – it makes a concrete difference:

> The need for privileged fibres has an important consequence when one needs to consider group actions. A single automorphism of a principal bundle can be transported to its Lie groupoid and vice versa, using a chosen reference point, but for general (nontrivial) groups of automorphisms it is impossible to choose reference points consistently. The automorphism groups of a principal bundle and its Lie groupoid therefore do not correspond and there is a notion of group action for Lie groupoids which makes no sense for principal bundles. (Mackenzie, personal communication)[24]

In sum, a full analysis of the use of the term 'natural' by mathematicians through the ages would require a book-length treatment. As used today it possesses several shades of meaning, which blend into each other to some extent, relying as they do on a sense of freedom from arbitrariness and artificiality. Promoters of groupoids see them as a natural concept since:

(1) One comes across them in the course of carrying out research in many areas of mathematics, without resorting to artifice
(2) They embody a simple, non-artificial idea, which permits them to measure the symmetries of families of objects
(3) They permit one to model situations without requiring that arbitrary choices be made.

9.9 CONCLUSION

This chapter should be seen as an early foray into an extremely complex subject. We have discovered that mathematicians will on occasion argue for the acceptance and further study of a piece of theory by indicating a panoply of good qualities. Philosophers should note that in the case treated here arguments for the 'existence' of groupoids did not figure in the array surveyed. This is through no oversight on my part – mathematicians make no use of the idea in their advocacy of the groupoid concept. On the other hand, turning to the arguments they do use, it is reasonable to wonder why so many different types are employed. I would explain this by pointing out that individual mathematicians weight the candidate criteria for progress idiosyncratically. One sets greatest store by the unificatory power of a concept, another by the potential for applications, a third by its ability to help resolve outstanding problems. Brown, Weinstein and other promoters of groupoids may have their own preferences as to the

---

[24] See Hitchin (2001) for the 'democratic' advantages of groupoids.

reasons why groupoids should be accepted, which reflect their sense of what mathematicians should be aiming to achieve, but they wish to cater for as many tastes as possible. Something akin to what Laudan terms the *dominance* of a theory over another is at stake. For Laudan (1984), even if the aims of scientists are varied, a theoretical development will still be held unanimously to constitute progress when it satisfies each scientist's criteria.

I believe there is much we can learn from parallel work in science studies. For example, Jardine's *The Scenes of Inquiry* (Jardine 2000) presents a highly sophisticated pragmatist philosophy, which stresses the importance of what he terms 'calibration', the act of making a theory measure up to earlier ones. The considerations of importance treated in this chapter would seem to be mathematicians' forms of calibration, where solving old problems is just one form. A particular concern of this chapter has been a disagreement about the weighting of these forms of calibration. We have encountered the thought that, whether seen as a goal in itself or as a means to a further end, a bias against conceptual reformulation and development have acted to delay the acceptance of groupoids. From this perspective, it took the efforts of conceptually daring mathematicians, such as Grothendieck and Connes, to set up programmes in which the use of groupoids became a matter of course.[25] A still more daring act of faith has been needed to pursue higher-dimensional groupoids. This research has provided key insights into how to develop higher-dimensional algebra, whose revolutionary credentials we shall be considering in chapter 10.

It is worth noting that for debates about the value of groupoids even to begin, there must be some shared ground, one or more absolute pre-suppositions, held by the participants. We can detect one fairly clearly – the presupposition that there is a distinction between concepts of great mathematical importance and concepts which are mathematically point-less. People disagree about how this distinction fits with the space of mathematics, but do not question that it exists. Now, we could plausibly claim that the idea of such a distinction has been operating at least as far back as the Greeks. It is curious, then, that many contemporary varieties of philosophy of mathematics disregard it.

One vision I share with Lakatos about how the future of mathematics might be shaped involves encouraging both a heightening of the level of historical awareness among the mathematical community and a facilitation of the expression of critical attitudes. This is not to say that the present

[25] See Cartier (2001) for some insight into why groupoids link the geometric visions of these two giants of mathematics.

situation is hopeless, but it can only be good for the health of mathematics to make improvements. One might also hope for useful communication with those philosophers who are well-versed in areas of modern mathematics. Unfortunately, the latter are not too numerous at present, but successful exchanges should encourage an expansion. Eventually one might hope to construct arenas where mathematicians and those working in the history, philosophy and sociology of mathematics can come together to permit sustained discussion of common concerns to take place.

PART IV

The interpretation of mathematics

Up to this point in the book I have largely been occupied by what is some-times called *descriptive epistemology*, the philosophical analysis of the work-ings of a knowledge-acquiring practice. Philosophy in this vein searches out allusive examples of some discipline's conceptual, social and material apparatus at work. When Nancy Cartwright (1999b) studies the modelling of superconductivity she is not trying to clarify the conceptual presupposi-tions of this particular branch of physics *per se*, but rather discerning more general methodological principles. As I have stated already, one should not fall into the trap of demarcating this philosophical activity from some pristine form of normative epistemology. Normative notions may emerge from the description of valued pieces of scientific activity. For Cartwright the development of superconducting quantum interference devices, used in medicine to detect brain damage in stroke victims, is physics being done at its best. This vision leads her to advocate that more resources be devoted to the less glamorous, but more practical, areas of physics (1999a: 16–17).

But developments in a science may serve other philosophical purposes. Besides showing us how physics works, from time to time they change our outlook on fundamental concepts: life, time, causality, matter, the universe. Philosophers of science taking themselves to be the descendants of the Ionian Pre-Socratics have been drawn to the latest scientific developments. Currently, the obvious choice in physics is quantum gravity. It is clear that if ever a successful unification of general relativity and quantum mechanics takes place it will entail profound alterations to our physical understanding. My undertaking in this final part of the book is to do something more analogous to the work of contemporary philosophers of physics such as John Earman or Jeremy Butterfield on quantum gravity.

I think it important that some resources of the philosophy of real math-ematics be devoted to the analysis of research at the *coal-face*. Historical studies of the inauguration of novel practices from much earlier times are, of course, indispensable to a proper understanding of the field, but risk being perceived as an antiquarian exercise. It is necessary to think of them as having once been contemporary, a notion which can be encouraged only by portraying contemporary trends as historically achieved. Moreover, the study of ongoing research gives philosophy a current relevance, and in so doing, one is only following in the illustrious footsteps of Berkeley on the calculus and Russell and Frege on the foundations of arithmetic.

So, where in today's mathematics do you look for an equivalent of these seminal moments in the history of mathematics? Well, I believe I may have found an answer. It is certainly a risky game predicting what will be seen in a hundred years time as the most significant conceptual advances of the day in any given field. However, my bet, for what it is worth, is that the developments I shall discuss in the next chapter, which go by the name *higher-dimensional algebra*, will be included in any list of *philosophically* important mathematical innovations constructed a century hence. A suggestive piece of evidence that they are *mathematically* important is the fact that I shall touch on the work of half a dozen Fields' medallists. Without wishing to make any hard and fast distinction between mathematics and the philosophy of mathematics, I must, of course, also argue that philosophers should pay attention to the possible philosophical consequences of higher-dimensional algebra. Here is a selection:

(a) Many important constructions may profitably be seen as the 'categorification' of familiar constructions. Indeed, very important structures are reached quickly by categorifying simple structures such as the natural numbers or the integers.

(b) It provides a way of organising a considerable proportion of mathematics. It shows us that set theory talks about just one corner of the mathematical universe, where set theory is taken in its category theoretic sense, i.e., free of the unwanted structure provided by ϵ-trees, and offers the potential for a clearer idea of the structuralism operating within mathematics.

(c) These constructions have applications in mathematics, computer science, and physics. In particular, we hear that 'higher-dimensional algebra is the perfect language for topological quantum field theory' (Baez 2001a: 192).

(d) Higher-dimensional algebra blurs the distinction between topology and algebra. Pieces of algebraic notation are taken as dimensioned topological entities inhabiting a space. Deformation within that space then corresponds to calculation. In this way, higher-dimensional algebra accounts for many uses of diagrams as means with which to calculate and reason.

At the present time it is difficult to gauge the size of the transformation underfoot. There are a whole range of definitions of higher-dimensional categories on the market, although for any dimension they are thought to be 'equivalent' at the next higher dimension. Of course, I am open to the charges that the theories I am occupying myself with have not taken on sufficiently definitive a form to allow for proper philosophical scrutiny, and that others deserve earlier treatment. Fortunately, philosophers of quantum gravity have had to defend themselves against similar charges and their defence has been performed very ably in a way that can be translated to my field (cf. Butterfield and Isham 2001, sec. 1.2).

When material is explained for a case study, there is little expectation that more than a handful of philosophers will work on it, except perhaps to refine or to challenge the morals being drawn from it. Quantum mechanics and general relativity have played a different role in the philosophy of physics. Philosophers come back to them time and again, as they are increasingly doing to quantum field theory and will eventually do, one imagines, to quantum gravity. If you want to be a philosopher of physics, you simply must be conversant with these theories. On the other hand, the prospective philosopher of mathematics quickly gathers that some arithmetic, logic and a smattering of set theory is enough to allow her to ply her trade, and will take some convincing that investing the time in non-commutative geometry or higher-dimensional algebra is worthwhile. One of the main purposes of the book has been to argue against this.

Imagine that twentieth-century philosophy of physics had run differently and that neither quantum mechanics nor general relativity had received much treatment from that quarter. One day some bright spark decides to write a book whose last chapter outlines the state of quantum gravity. I find myself now in a similar situation. Necessarily, therefore, a fairly large part of chapter 10 is expository and insufficiently philosophically 'processed'. For this I can only ask for the reader's forbearance, but I may be able to sweeten the pill with the observation that if my prediction that higher-dimensional algebra will emerge as a key mathematical theory is correct, then it becomes a topic about which philosophers of mathematics will have to inform themselves. Remember that Bertrand Russell once had to convince himself that it would pay to learn what Weierstrass, Dedekind, Hilbert, Frege and Peano were talking about.

CHAPTER 10

Higher-dimensional algebra[1]

Mathematical diagrams may well have been the first diagrams. The diagram is not a representation of something else; it is the thing itself. It is not like a representation of a building, it is like a building, acted upon and constructed.

(Netz 1999: 60)

Angular momentum and the topology of knots and links are a fantasy and fugue on the theme of pattern in a formal plane. The plane sings its song of distinction, unfolding into complex topological and quantum mechanical structures.

(Kauffman 1991: 621)

10.1 INTRODUCTION

Mathematicians are in the business of interpretation. Their lives are spent on open-ended quests for improved reformulations and reconceptualisations, where the familiar and taken-for-granted may at any moment be cast in a surprising light. The real numbers are seen at one time as the natural completion of the set of rational numbers, and later as just one such completion – the completion 'at infinity' – alongside infinitely many p-adic completions for prime p. The Euler characteristic is first seen as a regularity holding between the number of vertices, edges and faces of a polyhedron, later understood to be a topological invariant of a triangulable space of any dimension, and is now also seen by geometric probability theorists as a valuation on the algebra of sets generated by polytopes in R^n.

Consider the following spaces: the Eilenberg–Mac Lane space, K(Z, 2), whose only non-vanishing homotopy group is $\pi_2(X) = Z$; the space of unit

---

[1] I have met philosophers of mathematics on the practice-oriented side who from reading Lakatos believe there to be little in the way of informal mathematical exposition available. I can only encourage them to look harder. One mathematician who deserves lavish praise for the breadth and quality of his exposition is John Baez at the University of California, Riverside (http://math.ucr.edu/home/baez). I owe much of my understanding of the material in this chapter to his efforts and to comments in correspondence with him. There is no better way into the topic of higher-dimensional algebra, especially its relevance to physics, than through his papers and informal writings.

vectors modulo phase in a Hilbert space of countable dimension; CP^∞, the direct limit of finite dimension complex projective spaces, CP^n; the classifying space for principal $U(1)$ bundles; the complex vector space of non-zero rational complex functions in one variable modulo constants; and the space of configurations of integer-labelled points on the surface of a sphere, whose labels sum to zero. Now, in a certain strong sense, namely up to homotopy, these are just different descriptions of the same space. Each way of thinking about this space is valuable; each presents the prospect for novel connections. We know there is an Eilenberg–Mac Lane space $K(Z, 3)$, whose only non-trivial homotopy lies in dimension 3. We also know alternative descriptions of it: integer-labelled particles on the 3-sphere, etc. But in view of the consideration that it is potentially important for quantum mechanics, coming fourth in line after Z, $U(1)$, and the space of pure states in Hilbert space, finding some new ways to look at it might well be useful for physics.[2]

The fact that the philosophy of mathematics appears largely to have forgotten about this interpretative side of doing mathematics will, I feel, appear strange to our descendants. After all, the philosophies of Frege, Russell and the early Wittgenstein originated in a maelstrom of competing logical reinterpretations of mathematics (Ferreirós 2001). They arose from attempts to bring some order to the tumultuous nineteenth century. Looking back to the end of that century, we see mathematicians and philosophers beginning to perceive the opportunities opened up by the conceptual reinterpretations that had taken place within mathematics during the previous decades. No more important lesson was being learnt than that one should not worry unduly about the circumstances in which a mathematical entity of interest came to prominence. As we saw in chapter 4, just because you are dealing with a collection of functions defined on a complex curve does not mean that you should not think of them as possessing deep similarities to a collection of algebraic numbers generated by extending the rational numbers. By freeing these collections from the mathematical context of their discovery, they could both be seen simply as fields and properties pertaining to both could be discovered and justified at one stroke.

By the 1920s, the extraordinary result had been established that this process could be taken to such an extreme that any algebraic, analytic or geometric entity, any collection of such entities, and any mapping between collections of such entities could be seen as the same kind of thing – a set. This finding cast a long shadow over the philosophy of mathematics for the

---

[2] See 'week 149' at Baez's website (Baez 2001b).

rest of the century. If every mathematical entity is representable as the same kind of thing, the train of thought went, for the purposes of philosophy we need not concern ourselves with what is special about entities which are considered as possessing, say, geometric features or combinatorial features.

Mathematicians, on the other hand, recovered from the surprising universality of set theory. They benefited from the associated freedom to impose what could appear on the surface to be unnatural structures on given collections of entities. Objects could be examined in novel and illuminating ways, as when, as we saw in chapter 4, Marshall Stone defined a topology on the very algebraic concept of the set of prime ideals of a ring. But, of course, there were limits to this process. Penetrating interpretations do not grow on trees, but rather emerge through hard thinking about particular problems. We saw that Stone's construction owed much to Dedekind and Weber's remarkable idea to put into correspondence the prime ideals of a ring of functions and the points of an algebraic curve, so we certainly should not credit set theory unduly. While it is expressive enough to be able to represent more or less any desired construction, it has a problem in that it does not know how to say 'No'. It cannot distinguish between those constructions that the mathematically literate will realise are patently pointless and those that stand at least some chance of gainful employment. The question arises therefore as to whether we can say anything about the boundary which separates the potentially useful from the patently pointless.

At first glance, one might think that the motley of concepts and techniques used by mathematicians is so unruly that we could not hope for anything more than set theory's capacity to embrace them all, however indiscriminately. But, as you will already have glimpsed from this book, one way of reimposing some order has been the use of category theory. Category theory allows you to work on structures without the need first to pulverise them into set theoretic dust. To give an analogy in the field of architecture, when studying Notre Dame cathedral in Paris, you try to understand how the building relates to other cathedrals of the day, and then to earlier and later cathedrals, and other kinds of ecclesiastical building. What you don't do is begin by imagining it reduced to a pile of mineral fragments. Similarly, when category theorists think about those very important entities known as Lie groups, smooth manifolds that behave like groups, they do so by looking for group objects in the category of smooth manifolds, rather than by looking for sets which happen to be manifolds with compatible group structure. The idea here is to work out a theory of groups and then see which categories can support it, rather than just start out with sets:

Out of laziness and ignorance, people usually work in the category of sets as a kind of 'default setting'. This category has many wonderful features – it's like a machine that chops, slices, dices, grates, liquefies and purees – but usually you don't need *all* these features to carry out a particular task. So, one job of a category theorist is to figure out what features are actually needed in a given situation, and isolate the kind of category that has those features. (Baez 2001b)

It certainly could not be said that the story of category theory has been fully told today, not at any rate in the philosophical literature, but what I shall be describing here is an even grander project known as higher-dimensional algebra or higher-dimensional category theory, which pictures sets as 0-categories, categories as 1-categories and then continues up the dimension ladder, ultimately all the way to ω. We look first at categorification, the process of ascending this ladder (section 10.1). Next we see how higher-dimensional algebra gives us a way of organising tracts of mathematics. It transpires that some mathematical constructions which are deemed very important by parts of the mathematical community may be reached in a very few steps from the most basic operations when viewed in an n-category theoretic light, while a set theoretic perspective provides no insight into their importance. A section devoted to new insights into the diagrammatic basis of some forms of notation follows. Finally, we turn to look at applications in computer science and physics.

10.2 CATEGORIFICATION: MAKING DISTINCTIONS

When we consider a given finite set, all we can say of any two of its elements is whether or not they are equal. We need additional structural resources to be able to undertake a more refined analysis, when, for instance, we want to say that two things are different, yet similar in some sense. We might do this by moving up a level and comparing two sets. Now the sets may be different, and yet similar to the extent that there is a bijection between them. This similarity is made plain by the mapping which takes a set to its cardinality. Equinumerable finite sets are sent to the same natural number. This forgetting of extraneous information is fine if, for example, we merely wish to know whether the number of forks on a table is equal to the number of knives. But reducing each set of cutlery to its cardinality may lose information of interest to us. Perhaps we cared about there being a particular bijection which allowed us to lay out matched sets from the motley collection in our cutlery drawer for a dinner party. Sometimes less information is a good thing, sometimes not.

Mathematically, the passage from finite sets to natural numbers may be described as *decategorification*. In the category of finite sets, isomorphic objects have been identified and the mappings forgotten. The reverse process is known as *categorification*. We see these processes occurring in combinatorics. Stanley (1999: 219–29) lists sixty-six different combinatorial interpretations of the Catalan numbers, defined for $n \geq 0$ as $c(n) = \binom{2n}{n}/(n+1)$, i.e., $1, 1, 2, 5, 14, 42, 132, \ldots$ As an exercise, he urges the reader to demonstrate the validity of all these interpretations, ideally by finding 4290 'simple and elegant' bijections between the various sets being counted, i.e., one for each ordered pair, rather than merely equating each to the Catalan numbers.

Let us give an example of this. The number of binary rooted trees with n vertices and the number of ways of inserting n pairs of brackets into $(n+1)$ items in pairs are both equal to the nth Catalan number. For instance, for $n = 3$, we have 5 bracketings

$$(a(b(cd))), ((a(bc))d), ((ab)(cd)), (a((bc)d)), (((ab)c)d),$$

and 5 trees

I have ordered these sets to make obvious a 'simple and elegant' bijection.

In modern combinatorics we would not talk here about isomorphisms between sets, but rather of *species*. Species allow us a way of talking about labelled combinatorial objects while committing to no definitive assignment of labels and, so Rota forcefully argues in his preface to Bergeron *et al.* 1997, should revolutionise the field of combinatorics. They permit a rigorous understanding of the apparent magic of the reasoning which reveals the number of binary rooted trees possessing a given number of nodes as the corresponding Catalan numbers. According to this reasoning, a binary rooted tree is either empty or composed of a root with two further (possibly empty) binary rooted trees attached. This may be symbolised as the relation:

$$B = 1 + B \cdot X \cdot B = 1 + X \cdot B^2$$

Then,

$$B(x) = 1 + x B^2(x).$$

Solving this quadratic,

$$B(x) = \frac{1 - \sqrt{1 - 4x}}{2x},$$

and so expanding this,

$$B(x) = 1 + x + 2x^2 + 5x^3 + 14x^4 + 42x^5 + 132x^6 + \cdots$$

Reading off the relevant coefficient gives us the number of binary rooted trees with a given number of nodes.

When we speak of equivalence of species, we are invoking the idea of equivalence in categories other than that of finite sets, and hence require a broader way of understanding the not-identical-but-equivalent within any category. To this end, we define a pair of objects in a category to be isomorphic if there are two arrows going in different directions between them whose composites are the identity arrows for the objects. What the process of decategorification amounts to is the forming of equivalence classes of isomorphic objects. It may be applied to any category. Applied to the category of finite sets and mappings it yields the natural numbers. Applied to the category of finite dimensional vector spaces it again yields the natural numbers, the dimensions of the spaces. We can then make sense of the move from Betti numbers to homology groups in the history of algebraic topology, discussed in chapter 7, as an example of the opposite process, *categorification*, here from the natural numbers to vector spaces. The great advantage about working in this more structured setting is that one now has algebraic information on the mappings between topological spaces in terms of how they are represented by morphisms between vector spaces. Instead of merely assigning the number 2 to the surface of a doughnut to represent how it may be cut along two circles to form a rectangle, we can now assign to it the vector space R^2 and then represent mappings between such doughnut surfaces in terms of vector space transformations.

When you start to think about the processes of categorification and decategorification, you realise they are rife. Recall from chapter 9 the fundamental groupoid of a space, Y, with base points every point of Y. This is the category where the objects are the points of Y, and the arrows between x and y are homotopy equivalence classes of paths from x to y. What does the decategorification of this category look like? Well x and y will be counted as isomorphic if there are paths running between them in opposite directions such that the composite paths are homotopic to staying put at each respective point. All this requires is the existence of a single path between them, for then running along it from x to y and then returning in reverse is homotopy equivalent to staying at x, and similarly for y. The object x finds itself in the isomorphism class of all points of Y to which it is path connected. By decategorifying to the set of path components, we have passed from *how* points are connected to *whether* they are.

The following rather more involved examples show that some of the extra structure of a category may be preserved in the decategorified set: the character ring of a group is a decategorified version of its category of representations; the second cohomology group of a space, X, is the decategorification of the category of complex line bundles on X. Fortunately, we can also see this kind of preservation in the familiar example of the category of finite sets. This category has sums and products, which correspond to disjoint union and Cartesian product, respectively. Traces of these are preserved in the usual addition and multiplication operations within the set of natural numbers.

The homotopy theoretic example above is very useful to help us develop these ideas. We have seen how we must not treat two objects of a category as equal when they are merely isomorphic by learning how to refine the notion of sameness. However, we might wish to distinguish not only between *objects* being identical or equivalent, but also *arrows*. After all, going from Bangor to London then retracing your steps really does not seem to be the same as spending that time pottering about your garden in Bangor. At the level of categories we are forced to choose between declaring them to be the same or different, but it would be good to have the resources to call them different but equivalent. To be able to do this what we would need is a collection of 2-arrows, each going from one ordinary arrow to another, and satisfying suitable conditions. We should be able to compose compatible 2-arrows with an associative operation, and there should be identity 2-arrows for each arrow. Then we could say that two ordinary 1-arrows, f and g, between objects A and B are equivalent if there exist a pair of 2-arrows, α from f to g and β from g to f, such that $\alpha \cdot \beta$ is the identity 2-arrow on g and $\beta \cdot \alpha$ is the identity 2-arrow on f.

Homotopy theorists were driven up the dimension ladder with the realisation that n-groupoids model homotopy n-types. An n-groupoid is an n-category all of whose morphisms have 'inverses' up to an equivalence at the next level, up to an equivalence two levels above, and so on up to equivalence at the top level. A homotopy n-type is a homotopy class of spaces all of whose homotopy groups vanish above the nth. If mathematicians could work out how to give some sense to the notion of infinite-dimensional groupoids, they would then have algebraically circumscribed this field.

Let us take a look at the first steps up this ladder in terms of the surface of our planet. Now, any path from the North Pole to the South Pole may be continuously deformed into any other. In other words, from the fundamental 1-groupoid perspective, between any two objects there is a single arrow. But this does not appear to have captured what it is like to live on the Earth. The path between the Poles down the Greenwich meridian seems worth distinguishing from the way down the International date line. Let us keep the paths distinct and introduce 2-arrows. These can correspond to ways of sweeping a path from itself to another path. For example, we can sweep the Greenwich meridian at a constant rate eastward onto the date line. But we would not want to count this sweeping as different from a broadly similar sweeping which tarried a while as it passed down the longitude of, say, New Delhi. On the other hand, we would want to count it as different from the sweeping which proceeds westward at the same rate as well as from the sweeping proceeding eastward at three times the rate, which covers the surface of the Earth one and a half times. As 2-arrows between paths, then, we have homotopy equivalence classes of paths between paths. You may be able to see that these classes may be put into correspondence with the integers. Our original meridians are now not the same, but they are equivalent, since sweeping eastward by half a revolution followed by westward by half a revolution is equivalent to forcing the meridian to stay put.

The construction I have just sketched, the fundamental 2-groupoid, has detected more about the homotopy of the surface of a sphere, but there seems no reason to stop at the 2-level. Indeed, since there is no n above which the 2-sphere has zero homotopy, a fundamental ω-groupoid would be needed to capture its full homotopy. We seem to be moving towards a theory of 2-categories which have categories between their objects, i.e., $\hom(U, V)$ is a category for any pair of objects, and from there to n-categories for any n, having $(n-1)$-categories between two objects. This is precisely the theory of strict n-categories. But it is an approach which does not meet mathematicians' needs very well. At the level of 1-categories, either a composition along two sides of a triangle is the same as the third side or it is different, but at the level of 2-categories we have a chance to weaken equations governing the 1-arrows. The homotopy theorists were here first with their weakened associative law.

I have not been very precise so far in my mention of paths, so we shall have to enter into a little more detail to see why we need to weaken. A path between two points, x and y, of a space X is defined as a mapping, f, of the interval $[0, 1]$ into X, such that $f(0) = x$ and $f(1) = y$. Think of the path being run in a unit of time. But then what do we mean by the composition of f with another path g from y to z? If we join them the obvious way it will take us two units of time to get from x to z, but then strictly speaking it will not form a path. So what we do is run this composite path at double speed, e.g., at time ½ we are at y. What happens if we compose again with a third path, h, from z to w? I hope you can see that this will involve running f and g at quadruple speed then h at double speed. If, however, we compose in the other order, the composite of f with the composite of g and h, the resulting path runs along f at double speed and then g followed by h at quadruple speed. Now these paths are not identical, but they are equivalent. At the level of the fundamental 2-groupoid, we can say that the composite arrows are equivalent. Indeed, we can even designate a specific 2-arrow to do the job. The following diagram should be sufficiently suggestive of what is involved:

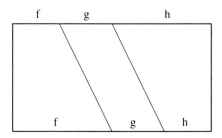

Weak 2-categories have the property that the associativity law for 1-arrows only holds up to isomorphism at the next level. These are also known as bicategories as distinguished from (strict) 2-categories. Since the weak ones are the ones we meet 'in nature' I shall just call them 2-categories. An example of their use is the way they can be used to express what is known as Morita equivalence for algebra-like objects (see appendix).

Then we can advance up the dimension ladder to 3-categories, and so on. In a 3-category, the equations governing the associators of 2-categories become isomorphisms at the level of 3-arrows. Complicated sets of coherence conditions must now be written down. The problem is to find a principled way of doing so. As Baez and Dolan put it:

To advance further in n-category theory, it is urgent to define 'weak n-categories' for all n. It is clear that new ideas are needed to do so without a combinatorial explosion, since already the explicit definition of a tricategory takes 6 pages, and that of a triequivalence 13 pages! (1995: 19)

Finding a tractable definition that copes with subtle shades of sameness and difference is no easy matter. Many proposals are on offer involving spheres, cubes, simplices, and other shapes in their constructions. Leinster (2002) has provided an up-to-date list of a dozen varieties of n-category, which in some sense are hoped to be equivalent. There is still much to play for.

Let us now return to categorification from the 0- to the 1-level, that is, from sets to categories. Monoids are a prevalent, basic kind of structure. They are defined as sets possessing an associative binary operation with a unit element. To categorify this notion we shall need a multiplication on objects, so that the object $A \times B$ exists for all A and B. We could then categorify the associativity law $(ab)c = a(bc)$ strictly to require some associative product on objects $(A \times B) \times C = A \times (B \times C)$, but for typical categories the relation holds up to isomorphism only. In the category of sets, for example, with multiplication corresponding to some specified cartesian product, there is an obvious isomorphism between these products. For each triple A, B, C we can specify a 1-arrow, an isomorphism known as the associator $\alpha_{A,B,C}$. These associators satisfy a similar equation to the one satisfied by the 1-arrow associators in a 2-category. So objects in a category like Set which has a product behave like 1-arrows in a 2-category, and 1-arrows like 2-arrows. We seemed to have slipped a dimension. This suggests an important way of generating categories with more structure. We take an n-category with one object, X say, one 1-morphism 1_X, and so on up to one $(k-1)$-morphism. Then we reindex the remaining morphisms so that the k-morphisms are taken as 0-morphisms or objects, and so on. This produces the following family.

Table of k-*tuply monoidal* n-*categories*

	$n = 0$	$n = 1$	$n = 2$	. . .
$k = 0$	sets	categories	2-categories	
$k = 1$	monoids	monoidal categories	monoidal 2-categories	
$k = 2$	commutative monoids	braided monoidal categories	braided monoidal 2-categories	
$k = 3$	" "	symmetric monoidal categories	weakly involutory monoidal 2-categories	
$k = 4$	" "	" "	strongly involutory monoidal 2-categories	
$k = 5$	" "	" "	" "	
. . .				

Source: Baez and Dolan (1999).

There are various ways of moving about the table. *Decategorification* takes you in a westerly direction from (n, k) to $(n - 1, k)$, as when we passed from a fundamental groupoid $(1, 0)$ to the set of path components $(0, 0)$. *Forming the centre*, a generalisation of the algebraic operation which gives you the elements of a group which commute with all other elements, takes you southerly from (n, k) to $(n, k + 1)$. *Looping*, isolating one object of the category and reindexing j-morphisms as $(j - 1)$-morphisms, allows you to pass south-westerly from (n, k) to $(n - 1, k + 1)$. For example, monoidal categories arise from looping 2-categories. *Stabilisation* involves a southerly movement from (n, k) to $(n, k + 1)$, resembling the process in homotopy theory of suspension eventually leading to stable homotopy groups as we saw in chapter 3. Many of these ways of moving, or functors, have a kind of inverse, known as an *adjoint*, in the reverse direction (see appendix).

Now, one of the main ideas is that starting from some very basic constructions – natural numbers, integers, and so on – and applying these functors we end up at some important mathematics. For instance, infinitely categorifying and stabilising the integers give you what is called the sphere spectrum, something central to algebraic topology. This programme evidently has enormous scope:

It is clear . . . that the set-based mathematics we know and love is just the tip of an immense iceberg of n-categorical, and ultimately ω-categorical, mathematics. The prospect of exploring this huge body of new mathematics is both exhilarating and daunting. (Baez and Dolan 1999: 32)

'Daunting' indeed – the path is still littered with huge conceptual challenges. In section 10.3 we shall see what sense we can make of parts of mathematics with this table.

10.3 HIGHER-DIMENSIONAL ALGEBRA AS AN ORGANISING LANGUAGE

In his account of what is at the heart of the so-called *Science Wars*, Hacking contrasts Kuhn's vision of science proceeding through large jumps with the physicist Steven Weinberg's much more gradualist account. Weinberg claimed that 'as far as culture or philosophy is concerned the difference between Newton's and Einstein's theories of gravitation, or between classical and quantum mechanics is immaterial' (Hacking 1999: 89). This may seem like the very essence of nonsense to philosophers of science, whose discipline has been described as emerging from these two great convulsions in early twentieth-century physics, but Weinberg is not alone here. The mathematical physicist Rovelli makes a similar point:

In my opinion, the emphasis on the incommensurability between theories has probably clarified an important aspect of science, but risks to obscure [*sic*] something of the internal logic according to which, historically, physics finds knowledge. There is a subtle, but definite, cumulative aspect in the progress of physics, which goes far beyond the growth of validity and precision of the empirical content of the theories. In moving from a theory to the theory that supersedes it, we do not save just the verified empirical content of the old theory, but more. This 'more' is a central concern for good physics. It is the source, I think, of the spectacular and undeniable predicting power of theoretical physics. (Rovelli 2001: 116)

This preservation of the insight of the old theory leads Rovelli to the conclusion that scientists such as Copernicus, Kepler, Einstein and Dirac are better seen as conservatives rather than as revolutionaries. So that:

figuring out where the true insights are and finding the way of making them work together is the work of fundamental physics. This work is grounded on the *confidence* in the old theories, not on random search of new ones. (Rovelli 2001: 117)

This is very much in line with the ideas of the philosopher of science Michael Friedman (2001: 60).

Now, philosophers of mathematics gazing back through time have found it difficult to see anything other than a hiatus marked by the advent of logical rigour brought about by the foundational crisis, the *foundationalist filter* I spoke of in chapter 1. But could we not say something plausible about mathematics, resembling Weinberg and Rovelli's positions, by playing down this perceived cleft between the nineteenth and twentieth centuries? Well, listen to Sir Michael Atiyah:

Whereas nineteenth century mathematics was primarily concerned with functions of one variable, the dominant theme of the twentieth century has been the problem of many variables. Great emphasis has therefore been put on basic structural features and these have in turn led to spectacular links between the discrete and continuous aspects of algebraic equations. (Atiyah 1976: 299)

Atiyah's idea is based on the observation that while the line can be folded up in just one way, and the plane into closed Riemann surfaces, we still do not know all the ways three-dimensional space can be folded up. The principal reason topology becomes so prevalent in the twentieth century is not now the search for a rigorous foundation for general spaces, but rather because the natural direction for mathematics was the study of equations in several variables whose more intricately structured solution spaces required the resources of a robust theory of space. As we saw in chapter 7, on the eve of the twentieth century this was one of the reasons Poincaré gave when explaining why he felt required to develop *Analysis Situs* for higher-dimensional spaces.

This depiction of what changes between the last two centuries allows Atiyah to propose a continuity to mathematics:

> modern mathematics is not as divorced from traditional mathematics as is sometimes implied. Mathematicians have regrouped their forces and spread out in different directions but the basic objectives are still the same. (Atiyah 1977: 73)

Some will, no doubt, find this rather questionable, but we should surely give it due consideration. There are different levels of absolute presuppositions, as Collingwood puts it. Perhaps some of these have changed very little over long stretches of time. Let's run with Atiyah's idea a little further. If the twentieth century marks the advent of the kind of structural thinking which category theory so often captures very well, brought about by the necessity of studying higher-dimensional spaces algebraically, analytically, topologically and geometrically, it would be apt if the twenty-first century was governed by structural consideration at the level of higher-dimensional categories. In other words, perhaps we have a scheme:

Nineteenth century	The study of functions of one variable
Twentieth century	The study of functions of many variables
	The search for structural features captured by 1-category theory
Twenty-first century	The search for structural features captured by n-category theory

This middle association is suggested by the fact that category theory is not apparent in the more elementary parts of mathematics, and begins to emerge only during the move to higher dimensions, as our examples of the section 10.2 indicated. We can see another instance of this when on higher-dimensional spaces the fundamental theorem of calculus becomes a result in differential geometry:

> Every time we integrate a function $\int_a^b f(x)dx$ we are concerned with a 1-form $f(x)dx$ on an interval $[a, b]$ with values in the Lie algebra of real numbers, and the integral is an element of the Lie group of real numbers. – common room conversation. (Sharpe 1997: 95)

Sharpe reinterprets the theorem as a result about the Darboux derivative where values are taken in a Lie group other than the reals. Category theory now starts to play a part, e.g., the functor from Lie groups to Lie algebras, categories of representations, etc. Today, at the beginning of the twenty-first century, mathematicians are finding they need higher-dimensional categories of representations of higher-dimensional algebraic objects, Lie 2-algebras, and so on.

Further support for the middle association comes from the fact noted in last chapter 9 that groupoids emerged in 1926 when Brandt extended to four variables Gauss's work on quadratic forms in two variables. This shift from group to groupoid takes us from the zeroth to the first column of the category table.

No doubt this is a terribly oversimplified picture, but even so I think it is interesting merely at the level of providing an alternative vision. The advice it suggests for philosophers is that rather than see the explosion of new mathematics as an unruly mess and fall back to the position that at least set theory can cope with more or less everything, we try instead to depict what there is more closely. The fact that mathematicians can see analogies between the mighty Langlands programme and topological quantum field theory (Kapranov 1995) should encourage us. If there is a modicum of truth to the picture, it can also provide us with a way to answer those who believe that the philosophical need for case study material can be met without straying beyond, say, 1930. In chapter 4, I presented the Dedekind–Weber paper of 1882 as marking a watershed in the development of structural thinking. One can see the development of their analogy right up to Weil and beyond as operating at the same level of abstraction, even if the ideas become recursively richer.[3] What seems to be happening with higher-dimensional algebra is the generation of a higher level of abstraction.

If we return to our table of n-categories, we can see that it does throw some light on the matter. First, we need to mention that the table can appear in three varieties: with nothing added, with inverses (up to equivalence), and with duals. Inverses we have met in our discussion of groupoids. Duals may be motivated by the notion of adjointness in Hilbert space theory and category theory. Linear Hilbert space operators $F: H \to H'$ and $F^*: H' \to H$ are adjoint if $<F\varphi, \psi> = <\varphi, F^*\psi>$, for all vectors in H and H'. F and F^* need not be inverse operators, but are in some sense an approximation. In the appendix the reader may see how closely the definition of adjoint functors resembles this one. We also need the notion of 'freeness' to continue. Mathematical entities are said to be free when they have all the properties they are required to, but nothing more. For example, the

---

[3] Thanks to José Ferreirós for that thought.

free (commutative) group on one object is forced to have an element with an inverse, and the means to compose associatively however many copies of the element and its inverse using only cancellation of inverses. It is thus isomorphic to the integers.

Let's now focus on the case where the categories come with duals as it is in some sense the most interesting. In general, something living in the (n, k) place of the table may be thought of as a collection of n-dimensional things living in an $(n + k)$-dimensional world. Mathematicians and physicists are especially interested in what goes on in the $(1, 2)$, $(1, 3)$, and $(2, 2)$ positions. For example, Feynman diagrams live in the $(1, 3)$ position, labelled one-dimensional drawings in 4-space where because of the number of degrees of freedom line crossings they need not be marked as underpasses or overpasses. Free versions of categories in these positions come without the labelling.

We now encounter the 'tangle hypothesis' which states that framed oriented n-tangles in $n + k$ dimensions are the n-morphisms of the free k-tuply monoidal n-category with duals on one object. Let us run through the situation for the $(1, 2)$ position where we are concerned with 1-dimensional things living in a three-dimensional world. Here the objects are collections of positively and negatively marked points. Ignoring the framing for the moment, an arrow is constituted by oriented lines going between marked points in the domain or target in specified ways. Essentially, we have oriented lines running in different directions and loops, possibly intricately tangled. Then the arrows from the object corresponding to the empty collection of points to itself are just oriented links. As a free type of entity, there will be a functor going from it to any other entity of that type we can find. In our case, any object in a braided monoidal category with duals will provide such a functor, and this functor will map a given knot to an arrow from the identity object to itself. Now, the category of representations of any quantum group is braided monoidal, and we are thus provided with a large supply of knot invariants.

What has happened here is in line with the usual practice of algebraic topology to engineer algebraic representations of topological situations. In other words, it has devised functors from some topological category to an algebraic one. Here, however, a new type of algebra is the target of the functor, something belonging to the field of what is now called *quantum* algebra. Something notable about this development is that whereas older invariants produce algebraic objects for each dimension, the objects of quantum algebra are tailored to the relevant dimension.

The case of the $(1, 2)$ position allows us an interpretation of what has been happening in the parallel field of *quantum* topology. Until the 1980s, knot invariants were unable to distinguish between knots and their mirror images. Vaughan Jones working on von Neumann algebras found a set of relations on generators which put him in mind of braids, entities closely related to knots. This permitted him a representation of knots sensitive to the two kinds of crossing there can be. The original invariants from topology all factored through the representation of spaces as n-groupoids, but braided monoidal n-categories with duals provide a richer environment to detect structure.

Higher-dimensional algebra can now provide us with the beginning of an answer to our question asked on p. 11 as to why Hopf algebras are worth studying while snooks probably are not. General Hopf algebras have representation categories which are monoidal with duals, while those which are *quasi-triangular* are important because their representation categories are braided monoidal with duals. These latter, also known as quantum groups, may be produced by deforming the universal enveloping algebra of semi-simple Lie algebras. The representational counterpart of this kind of deformation corresponds to a deformation of a symmetric monoidal category (1, 3) to a braided one (1, 2) within the 2-category of braided monoidal categories.

10.4 DIAGRAMMATIC CALCULATION

In spite of the very widespread use of diagrams by mathematicians to expedite their reasonings and calculations, philosophers have paid them scant attention. When they have done so it has often been to look on diagrammatic reasoning as a poor relation of verbal reasoning, quicker at times but dangerous, potentially able to offer new insights, yet not wholly reliable. Wasn't the nineteenth century one long exercise in showing us how our spatial intuition can lead us astray?

A minority position argues for pictorial reasoning to be seen as something stronger than this, something capable of reliably generating mathematical knowledge. Brown, for example, argues that Bolzano's picture of a continuous function allowed him to observe a result about continuous functions which later provided independent evidence that the arithmetisation of analysis was on track. Here we have an important role attributed to pictures:

Pictures are crucial. They provide the *independently-known-to-be-true* consequences that we use for testing the hypothesis of arithmetization. Trying to get along without them would be like trying to do theoretical physics without the benefit of experiments to test conjectures. (Brown 1999: 29)

So, the kind of diagram that produces an arithmetic identity is more than an inspiration. Like an experiment it provides independent access to the truth:

Some 'pictures' are not really pictures, but rather are windows to Plato's heaven. (Brown 1999: 39)

This idea of transparency is echoed in a superb book on complex analysis:

The basic philosophy of this book is that while it often takes more imagination and effort to find a picture than to do a calculation, the picture will always reward you by bringing you nearer to the Truth. (Needham 1998: 222)

Brown's philosophical position places the emphasis on what is to be known, for him the inhabitants of Plato's heaven, but it is equally important to think about the process of observation. After all, there is much one can understand about physics by considering the changing nature of instrumentation and experimentation. In modern physics, as Galison (1997) and others have described, the bodies of skills and knowledge necessary to maintain detectors of high-energy particles and extract observations from them are very richly structured indeed. So, similarly for mathematics, we find that diagrammatic technology has changed over the years bringing novel phenomena into view. The observational apparatus of today's mathematician resembles more the electron microscope or the radio telescope than it does the pane of glass constituting one of the 'windows to Plato's heaven'. What you see with it has much to do with its own nature, that is, what it allows you to do with it. Indeed, what we find when we bring contemporary varieties of diagrammatic apparatus into focus is that they cannot be neatly separated from what they might be thought to be about. Rather, attention shifts constantly between the probe and the probed.

The cases of diagrammatic reasoning considered by Brown suffer from lacking that element of systematicity that is so noticeable in, say, the use of algebraic notation in the hands of Fermat and Descartes. There a whole raft of problems could be solved in a uniform fashion. With the diagrams of Pythagorean pebble-arithmetic or Euclidean geometry, however, there remains a continual need for the invention of clever tricks. One-off flashes of inspiration are fine, but how much more interesting if behind the tricks we find something of more substantial calculus. As we shall now see, systematic, rigorous diagrammatic reasoning is becoming a reality.

But surely, one might have thought, the problem is not one of paying insufficient attention to the twentieth century. When you glance through a range of contemporary textbooks, a very low percentage have anything more than the odd diagram. And wasn't it *Bourbaki* who proudly ensured that so few diagrams appeared in their *Eléments*? Well, *Bourbaki* belong to what might be designated the high modernist period. Now, we inhabit a postmodern world, where:

> geometric representation theorems for the syntax of higher-dimensional categories are examples of what Street and others have begun calling 'post-modern algebra' ... In post-modern algebra, the abstract algebraic notions are very much alive, but the syntax of their operations is often represented by concrete structures drawn from outside algebra, typically geometric or topological in nature. (McIntyre and Trimble forthcoming: 11)

Whether or not we see anything in the designation beyond a joke, the signs coming from the front lines of mathematical research cannot be ignored. Spin networks, Feynman diagrams and spin foams, their higher-dimensional analogues, Penrose tensor diagrams, and even Charles Peirce's existential graphs all may be treated in this framework. These diagrams are not just there to illustrate, they are used to calculate and to prove results rigorously.

The possibility of doing so arises from the flip side of our facility to use algebraic objects to understand topology, as in the case described in section 10.3 of quantum group invariants of knot theory. Now, we work the other way around, using topological objects to allow us to calculate in algebra. For example, we can use framed tangles to help us calculate with representations of quantum groups. For a representation V with dual V^*, there are maps: $ev_v : V^* \otimes V \to \mathbf{C}$; $coev_v : \mathbf{C} \to V \otimes V^*$; $\psi_{v,v^*}: V \otimes V^* \to V^* \otimes V$, and its inverse, satisfying various relations. They may be represented diagrammatically as a cup, a cap, and two types of crossing, so that, for instance, the composite map $U_v = (ev_v \otimes id_v) \circ (\psi_{v,v^*} \otimes id_v) \circ (id_v \otimes coev_{v^*})$ is represented as on the left:

To help you to understand the representation, on the right I have shown the same diagram sliced through four times. Each of the three brackets represents a map between two slices.

The relations governing the maps are such that if we think of each curve of a diagram as a length of ribbon, any physical move we could perform on these ribbons keeping them flat in the plane corresponds to a legitimate algebraic calculation. We can then show that U_v has an inverse

and define this inverse in terms of ev, coev and ψ. We can also work out what the U map looks like for the tensor product of two representations $V \otimes W$:

This demonstrates the relation $U_{v \otimes w} = \psi^{-1}{}_{v,w} \circ \psi^{-1}{}_{w,v} \circ (U_v \otimes U_w)$. To write out the steps in the same linear form would be horrendous. There are eight cups, caps and crossings in total in the left-hand diagram, compared to three for U_v. Already in the algebraic manipulation corresponding to the first tangle transformation several pieces of rewriting would have occurred. It is not surprising then that a text book such as Majid (1995: 444–6) relies only on the diagrammatic proof.

Now there are two things to note here. First, without the diagrammatic representation it would be exceedingly hard to find a proof of this relation. Second, having followed a representation of the proof written in standard linear notation, one has no reason to be more confident in its correctness than having followed the pictures:

once the appropriate algebraic expression is found that represents such diagrams, it seems clear that algebraic computation will be no more easy than diagrammatic manipulations. (Carter *et al.* 1996: 65)

Matters become far more complicated than this when we try to repeat these constructions in the position (2, 2). Diagrammatic notation has shown itself easier to handle than the algebra in position (1, 2) and yet the four-dimensional algebra in position (2, 2) requires two-dimensional diagrams sitting in 4-space, or at least the projection of these surfaces into 3-space with all the accompanying kinds of intersection. While the latter are difficult to visualise and *a fortiori* to manipulate, the benefits of using spatial intuition over straight algebraic manipulation are such that it is worth developing ways to support this intuition even if this means we still cannot take in at a glance whole manoeuvres. We have three options:

(1) Project the knotted surface into 3-space marking points of intersection. (Instead of one type of intersection point with two types of crossing as in knot projections, we now have three types of point with a total of ten types of crossing.) Then find a set of legitimate moves on these projections.

(2) Present a *movie* of the surface – a projected surface via a series of stills, slices through the surface either side of singularities, each of which contains a link diagram. Legitimate moves then correspond to certain movie moves.

(3) Represent each still as a word written in symbols depicting caps, cups and crossovers, and then each movie as a sentence of such words. Legitimate moves then correspond to certain rewrite rules.

Option (3) exposes us to the risk of mistaking legitimate rewritings through the immense length of calculations, and makes them all but impossible to find; (1) may encode a calculation extremely efficiently, but presents the risk of the faulty use of our spatial intuition. Out of these options (2) with its combination of step-by-step procedures but reliance on 2D spatial capabilities seems the best suited to humans, especially if worked alongside (1). I can only urge you to consult Carter and Saito's beautifully produced book – *Knotted Surfaces and Their Diagrams* (Carter and Saito 1998) – to gain a proper sense of what is at stake. A projection of a surface translates to a movie of 13 stills, or to a sentence of over 400 letters. Rewriting the sentence representation, you have four pages of rules to follow.

Obviously at some point as we climb up the dimension ladder our intuition will fail us. But it is important to note that it is not just that this diagrammatic rendition makes the mathematics easier for us, more accessible to creatures with our capabilities, rather the fact that the algebra and topology are so intimately connected argues for the importance of the algebra as powerfully related to what we might call after Spencer-Brown the 'laws of form'. Hopf algebras, quantum groups and Lie algebras gain in naturalness, where snooks never will, for having this connection to topology.

The above is just one part of a large programme to use diagrammatic notation which shows:

(1) how certain algebraic relationships can be depicted and computed via diagrams;
(2) how diagrams lead to algebraic structures;
(3) how singular diagrams yield algebraic relationships;
(4) how diagrams can be used to anticipate certain algebraic structures.

(Carter *et al.* 1996: 2)

Regarding point (4), the discovery of algebraic objects, perhaps quasitri-
angular Hopf 2-algebras, whose representations form a braided monoidal
2-category with duals now becomes desirable to assist the development of
4-dimensional topology and physics.

Now remember that the proponents of this programme do not claim
that their diagrammatic reasoning cannot be fitted into some accepted
system such as set theory. Indeed, one may consult Joyal and Street (1991)
for a proof that some forms of it can be thus fitted, where they recast
the diagrams as topological subsets of the plane. But it's hard to escape
the feeling that something odd is going on here. Indeed, this form of
diagrammatic calculation has struck one of knot theory's leading exponents
as novel. The observation that one may reason about knots by manipulating
their projections has led Louis Kauffman to say:

Notationally the Jordan curve theorem is a fact about the plane upon which we
write. It is the fundamental underlying fact that makes the diagrammatics of knots
and links correspond to their mathematics. This is a remarkable situation – a
fundamental theorem of mathematics is the underpinning of a notation for that
same mathematics. (Kauffman 1991: 15)

This is an intriguing thought. The Jordan curve theorem tells us that:
*Any continuous simple closed curve in the plane, separates the plane into two
disjoint regions, the inside and the outside.* Reliance on it suggests that now
not only should we think of our notation devices as pieces of topology –
we need also think about the medium which allows the notation to be
inscribed upon it. Let's illustrate this idea by first moving down a dimension
to the more prosaic use of point-like notation. In the $n = 0$ column of the
table, in the $(0, 1)$ position we find that the free monoid on one object has
as elements finite sets of points inscribed along a line, with composition
equivalent to juxtaposition. It is equivalent to the natural numbers, **N**,
like Hilbert's strokes. Similarly, the free monoid with duals, i.e. groups, is
isomorphic to the integers. Integers are seen as classes of dots and duals
strung along a line, where an adjacent dot and dual may annihilate each
other, and such a pair may be created anywhere on the line.

By thinking about its topological character, we see the basis for traditional linearly written algebra. Imagine you are working with a monoid, that is, a set with an associative binary operation with identity. When you write down a term of the monoid, in effect you are labelling a string of 0-dimensional things sitting in a 1-dimensional world. Your ability to rewrite this term corresponds to a freedom to allow these labelled points to move along the line and compose in pairs. If you would prefer that they didn't interfere with each other quite so freely, you must put in brackets to stop them. If you do want associativity, on the other hand, then there is no need for brackets. You see this also in category theory, i.e., in position (1, 0). Here we have one-dimensional objects living in a one-dimensional world and we get associativity in the composition of three arrows for free.

Now, how do we know that when we write $a.b$ on a line that when we come back after a coffee break the a won't have slid past the b? Well, marks just don't move like that. Does the distance from a to b matter? No, it is purely a topological matter, and topologically the line with the point labelled b removed is such that a point on the left must stay there. Or more symmetrically, given that the points are not allowed to occupy the same point, the configuration space is a square without a diagonal, and so composed of two components. Restriction to a line thus prevents commutativity, just as brackets stopped associativity. But what if we didn't mind about the order of multiplication? Well, a representation of 0-dimensional things in a two- or higher dimensional world will allow this. We would be happy, for instance, to write down elements of a free commutative group and for them to move about with pairs of inverses annihilating each other, so that when we got back the calculation had been done. Had we left a camera running we would see a tangle traced out.

Exploiting the freedom of the page is precisely what the philosopher Charles Peirce did in his representation of propositional logic via the alpha portion of his existential graphs (EG). As we saw in chapter 2, for Peirce if A and B represent propositions, then inscribing them both on a page represents the assertion of their conjunction. This graphical notation uses the topology of the plane to show the irrelevance of order. We can see that it makes no difference whether A is to the right or left of B. Similarly, there is no need to mention associativity. So the six orderings of A, B and C each with two bracketings, that is twelve different syntactical forms, are represented by the same existential graph. An alpha graph works up to legitimate topological equivalence.

Peirce went on to construct the beta part of his graphs, which captures what was later called the predicate calculus. Here we use strings as well as letters and scrolls. Let's see how this works by representing the proposition 'No man has seen every city':

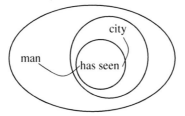

Peirce was aware that the use of logic presupposes a certain perceptual capacity. In 1902 he wrote:

Kant is entirely right in saying that, in drawing those consequences, the mathematician uses what, in geometry, is called a 'construction', or in general a diagram, or visual array of characters or lines . . . Thus, the necessary reasoning of mathematics is performed by means of observation and experiment, and its necessary character is due simply to the circumstance that the subject of this observation and experiment is a diagram of our own creation, the conditions of whose being we know all about.

But Kant, owing to the slightest development which formal logic had received in his time, and especially owing to his total ignorance of the logic of relatives, which throws brilliant light upon the whole of logic, fell into error in supposing that mathematical and philosophical necessary reasoning are distinguished by the circumstance that the former uses constructions. This is not true. All necessary reasoning whatsoever proceeds by constructions; and the only difference between mathematical and philosophical necessary deductions is that the latter are so excessively simple that the construction attracts no attention and is overlooked. (Ewald 2000: 635–6)

Quine also perceived that variables could be replaced by connecting lines. In a textbook he wrote on logic, he expresses the same sentence as follows:

( )(is a man.⊃ ~( )(is a city .⊃. has seen)

He continues:

But these 'quantification diagrams' are too cumbersome to recommend themselves as a practical notation. (Quine 1955: 70)

Peirce's system, on the other hand, has been described in much more favourable terms:

> if a facile and perspicuous notation is one that can be learned and easily manip-ulated, then years of experience with university students have convinced me that EG is the most perspicuous, and *Principia* notation the least. The unusual ease with which inferences can be drawn in EG is something of an unexpected bonus. (Roberts 1973: 126)

As you might have guessed, the beta system can be recast in terms of *n*-category theory. Each diagram may be seen as an arrow in a symmetric monoidal category, and with inference represented by 2-arrows, the typed predicate calculus is then reformulated as a monoidal 2-category (Brady and Trimble, unpublished).

The revival of diagrammatic calculi owes much to the efforts of Roger Penrose. He developed a kind of tensor notation where instead of indices one has strings:

> While these symbols look like tensors, *there is no summation on the indices*. The *identity of the indices is very important* – two symbols that share an index represent fragments of a diagram that are tied together at the site of this index. Abstract tensor algebra is a significant departure from other forms of abstract algebra. In the abstract tensor algebra there are symbols, separated by typographical distance, that are interconnected by their indices. (Kauffman 1991: 566)

'Typographical distance' is an interesting idea. For anyone who feels that the attempts of Field (1980) to spirit away mathematical structure into space-time regions and logic are somewhat suspect, it provides a possible key. Where Field quantifies over space-time regions with the formula $\exists x \, \exists y (Fx \; \& \; Fy \; \& \; x \neq y)$ to avoid the use of '2' in 'there are at least 2 Fs', mathematicians sense that this trick relies on making two distinguishable marks on a piece of paper, in this case x and y. In Peirce's system the twoness is quite evident:

Clearly much remains to be said on this topic, but let us now briefly turn to the question of why mathematicians have found it difficult to promulgate diagrammatic notation. In the late nineteenth century, Peirce was not alone in developing a notation for logic requiring two dimensions – Frege's *Begriffsschrift* took a clumsy tree-like form. In both cases problems with printing technology seem to have played a factor in their lack of success. This appears to be a general problem for diagrammatic techniques. Penrose says of his tensor notation:

Unfortunately the notation seems to be of value mainly for private calculations because it cannot be printed in the normal way. (Penrose and Rindler 1984: 425)

As printing technology has improved, two mathematicians deemed it necessary to remove doubts concerning its validity:

Penrose was the first to use the graphical notation for calculating with tensors. It is now currently used by theoretical physicists as a private device for quickly verifying complicated tensor formulas. A striking aspect of the notation is that it is pictorial rather than sequential or alphabetical. This made it difficult to print, which partly explains why no rigorous theory was developed. We believe that a notation which is useful in private must be given a public value and that it should be provided with a firm theoretical foundation. Furthermore printing techniques have improved drastically in recent years. (Joyal and Street 1991: 55)

'Drastically' perhaps, but drawing packages still do not allow physicists and mathematicians to construct diagrams as easily as they can write with Latex. Although mathematicians and physicists manage surprisingly well with ASCII when they discuss technical issues in newsgroups, some mathematicians I have consulted have said that there are things they have chosen not to publish because of the limitations of available drawing packages. Not everyone has the dedication of Carter and Saito, who may spend up to three hours on one of their diagrams (personal communication).

I hope that from this brief foray you will agree that research into the evolution of notational styles, and the ways printing skills and technology have interacted with notation formation through the ages, would be tremendously valuable. An important part of this research would concern differences between communities in their attitude towards notation. No doubt one would find great variation even within a single community, but perhaps the following comment might provide a starting point:

The notation used in physics is not designed to emphasize the logical relations of the concepts involved, but rather to facilitate explicit calculations. (Rabin 1995: 186)

Let us now turn to the physicists and others who might benefit from higher-dimensional algebra.

10.5 APPLICATIONS

We would expect, if it is half as important as I have been suggesting, that higher-dimensional algebra would find itself in a position to be able to provide novel 'constitutively a priori principles' for other sciences, to use the language of Friedman (2001). In other words, once in place it should enable a range of new empirical theories to become thinkable. In this section we shall see whether this may be so for quantum gravity. But before we head off along that path, I should first like to mention another potential area for the application of higher-dimensional algebra. I shall be brief here since this area, theoretical computer science, is still in a fairly precarious position. In principle the discipline has much going for it. Computer languages are formal languages, and programs written in those languages resemble proofs. One would think, therefore, that formal systems would be very useful in modelling programs and in the construction of new languages. However, the experience of some undergraduates passing through their degree is of an unbridgeable gulf between the highfalutin end and the practical concerns of computing. C++ and Java just do not seem to have been designed on the basis of a deep structural understanding of logic. Let us just for the moment, however, set aside these concerns and trust that there is a point to it all.

Foundational branches of mathematics have become a matter of practical interest in theoretical computer science (Taylor 1999). Logic becomes unrecognisable in the hands of 'informaticians', with a range of new logics, including the family of linear logics. The development of computer languages provides an excellent topic for philosophical reflection: the imperative/declarative dichotomy, object-oriented reasoning, concurrency, etc. Conceptual problems are prevalent, and several computer scientists have revisited Frege's writings to look for inspiration as to how to handle variables cleanly.

There needs to be philosophy of computer science separate from that part of philosophy of mind which concerns itself with artificial intelligence, and from the philosophical component of the work on technology in the sociology of science. It needs to create a space where various dualities can be examined – time/information, automaton/schedule, state/event, programme/context – which might find points of contact with philosophy of physics. Even at a fairly superficial level one can see connections between physics and computer science. Quantum field theory sees particles as quanta of fields, where the particles mediating the interaction themselves interact. Computer science is interested in modelling interaction between processes, seen themselves in terms of their interaction with an environment, where information 'tokens' trace out some path around a network. 'Process algebras' are employed to represent the binding together of processes into a composite with mutual communications internalised. As in field theory, interaction is local.

Rather like physicists of the 1930s finding a shortage of constitutive principles from mathematics to cope with quantum field theory, we hear a similar complaint today:

The notions of *function*, *set*, and *algorithm* were available 'off-the-shelf' from mathematics and logic for use in computer science. By contrast, there is no adequate pre-existing theory of processes, interaction, information flow etc. on which these 'second-generation' models can build. (Abramsky 1997: 1)

One notion which computer scientists have seen fit to construct to make up for this shortage is linear logic, a resource-sensitive logic devised by the proof theorist Girard. Semantics for linear logic include ones based on games, but intriguingly also ones based on our old friends Hopf algebras. And where Hopf algebras are to be found, you can be certain categories will follow.

It would take a monograph-length study to examine the ways in which computer scientists have invoked category theoretic constructions. Here, we must be brief. Later in the same article, Abramsky writes:

Traditional completeness theorems in logic have focussed on characterizing *provability* by validity in some model or class of models. The focus from the point of view of computation or proof theory is on the proofs themselves as mathematical objects, rather than on the mere fact of provability. The idea of full completeness is to have a model given by purely semantic means (i.e. independently of the syntax) such that every element of the model is the denotation of some proof (this can be stated in terms of the *fullness* of the functor from the free category based on the syntax, hence the name). (Abramsky 1999: 2)

Closely related is the idea of *full abstraction* for a programming language where a syntax-independent model provides the denotation for any programme. Notice how just as with the categorification from the set of path components to the fundamental groupoid, we have again a passage from *whether* to *how*, from whether *A* is provable from *B* to how it is provable.

Higher-dimension algebra can now come in to allow comparisons to be made between these models, but it appears elsewhere. In the theory of automata, automata are seen as different ways of taking an input of a certain kind and giving an output of another. The automaton may be described as doing this according to its internal state. What is needed is a language to say that two automata are 'different but equivalent', i.e., we need a notion of a 2-arrow from an automaton 1-arrow to another. Elsewhere, those modelling the syntax of a language deal with three levels: types, raw terms, and redux paths. It is not hard then to imagine a higher-dimensional syntax talking of maps between redux paths, and so on. In fact, people use algebraic topology to analyse computation in these languages for confluence.

I shall stop here in the hope that a pioneering philosopher or two might be inspired to move into this vast field, and turn instead to the even vaster field of mathematical physics.

Since the late 1970s there has been an enormous increase in the flow of ideas between mathematics and mathematical physics. Debts have been incurred not just by physicists, but also by mathematicians. One of these debts arises out of the response of the mathematical physicist Edward Witten to a problem raised by Michael Atiyah in 1988. As I have already mentioned, a few years earlier, after decades of little progress in producing more powerful knot invariants, a new polynomial, named after its discoverer Vaughan Jones, was found to be able to distinguish between some knots and their mirror image. Previous invariants had been unable to do this. Now, knots are one-dimensional things living in a three-dimensional space and yet Jones's invariant and its many successors were calculated from the knot's projection onto the plane, relying on the associated algebra being invariant under legitimate transformations of the projection. Atiyah wondered how to give a straight three-dimensional interpretation of these knot invariants.

Witten answered the challenge by constructing a quantum field theory on a three-dimensional manifold with boundary composed of surfaces marked by negatively and positively signed points. Embedded in the 3-manifold is an oriented tangle, whose open strings end at the marked points on the boundary. Witten then 'averages out the geometry' by performing an integration with respect to a measure on the space of connections (modulo gauge) on a principal bundle over the 3-manifold, yielding a topological invariant. Knot invariants arise from applying this integration to the case where a knot is embedded in the 3-sphere. His formula:

$$\int dA \exp[(ik/4\pi)S(M, A)] \, Tr \left(P \exp \left(\oint kA \right) \right),$$

is not rigorously defined because of the lack of a suitable measure, but it is possible to treat it heuristically to show that it does give the same answers as the Jones polynomial.

Atiyah subsequently extracted an axiomatic definition of a topological quantum field theory (TQFT) as a rigid symmetric monoidal functor from the category of n-cobordisms to the category of vector spaces. Rigid here means that it preserves duals. The TQFT is called unitary if the target category is the category of Hilbert spaces and compatible with a second sort of duality structure at the level of arrows.

A unitary TQFT in dimension n is a functor Z which assigns:

(1) A finite-dimensional complex vector space $Z(\Sigma)$ to each compact oriented smooth $(n-1)$-dimensional manifold Σ

(2) An operator between vector spaces $Z(\Sigma)$ and $Z(\Sigma')$ for each compact oriented n-dimensional manifold Y with boundary diffeomorphic to $\Sigma \cup \Sigma'^*$, where $*$ indicates reversed orientation.

Various axioms must be satisfied which are neatly encoded in Atiyah's category theoretic definition. Cobordism classes are used to count diffeomorphic space-times as the same.

One helpful set of associations to have in mind is as follows:

$(n-1)$-dimensional space	Hilbert space of states
Cobordism between two $(n-1)$-dimensional spaces (space-time)	Operator acting on states
Composition of cobordisms	Composition of operators
Identity cobordism	Identity operator
Time reversal	Adjoint operator (see Baez 2001a: 186)

Now there are three techniques used to construct TQFTs: one is exemplified by Witten's use of a path integral formulation and a Lagrangian; a second is category theoretic, using a description of cobordisms in terms of generators and relations; and, the third is combinatorial, about which more later. One interesting idea has been to extend the manifolds involved to allow corners. Now taking a time-slice through, say, a 2-cobordism produces a 1-manifold, which in turn may be seen as a cobordism between 0-manifolds. This is reminiscent of Carter and Saito's presentation of knotted surfaces. Where Atiyah gives his definition of a TQFT in terms of 2 levels, as implied by his use of categories, here we are dealing with the notion of an *extended* TQFT which rests on an n-category description. Extended TQFTs map from topological n-categories to hit algebraic n-categories. In other words, the algebra needs categorifying as well. Just as important objects are picked up by categorifying the integers, so interesting algebraic entities are generated starting from the complex numbers. One result of this process is that the notion of adjoint operators becomes categorified to adjoint functors. This is quite remarkable because the latter term was introduced several decades ago, based on what was thought to be a mere resemblance.

But why would anyone one want a TQFT, aside from making sense of the mathematics of knot invariants? Well, one of the central problems in reconciling quantum field theory and general relativity is that each has a failing. Quantum field theory is carried out on manifolds with a background metric in place. General relativity, on the other hand, is not a quantum theory. What everyone is after is 'a background-free quantum theory with local degrees of freedom propagating causally' (Baez 2001a: 178). However:

freedom in the construction of a (background independent) quantum theory of geometry is very limited. Thus, the mathematical structures, definitions and construction we use are . . . essentially unique. (Ashtekar 2002: 1)

To help with the search for these structures it was thought worthwhile to construct the third point of the triangle whose first two points are represented by general relativity and quantum field theory, by devising background-free quantum theories with no local degrees of freedom, i.e., TQFTs. These are not so unphysical as they sound. Indeed, classical general relativity in three dimensions has no local degrees of freedom, since the curvature is completely determined by the flow of energy and momentum. The four-dimensional general relativity we think relevant to our universe does, of course, have local degrees of freedom, but the expectation is that a four-dimensional TQFT will shed great light on it. Let's now look at the third way to produce a TQFT.

Particle physicists wishing to devise a theory of quantum gravity have tended to opt for string theory. By doing so they are staying close to the gauge field theories which have served them so well in the construction of the Standard Model. Relativists, on the other hand, have taken Einstein's theory as their starting point. Direct attempts to quantise general relativity have come unstuck because of the problem of forming Hilbert spaces from infinite dimensional spaces of connections modulo gauge equivalence, so one important step has been the reformulation of the theory in terms of novel variables, loops rather than points. Loop quantum gravity works by quantising holonomies along paths. The space of states for such a theory has been found to be spanned by what are known as spin networks.

Thus, loop quantum gravity theorists have revived a thirty-year-old theory of Roger Penrose. Around 1970, to avoid all the problems associated with continuum models, Penrose had constructed a combinatorial, discrete model of space. This involved merely a trivalent graph – three edges meet at each node – where each edge is labelled by a half-integer, corresponding to a representation of SU(2), the group involved in angular momentum in quantum theory. Penrose showed that Euclidean directions emerged from a large-scale spin network.

Intriguingly, the four-colour theorem can be recast in terms of a class of spin networks in which each edge is labelled '1' and which behave as though they can perform vector algebra. The theorem then becomes a result about solutions to a vector equation. This has not led to a simpler proof, but physical ideas may prove useful.

Recently, loop quantum gravity theorists have generalised spin networks to allow nodes of higher valency and labels corresponding to different groups and to different types of algebra. They have also moved up a dimension to model the dynamics of the system by what have been dubbed *spin foams*. Spin foam models of quantum gravity are analogues of Feynman diagrams, where path integrals are rewritten as a sum over spin foams connecting spin networks:

In spin foam models, the microscopic degrees of freedom are representations and intertwiners of the appropriate group (originally SU(2)) and live on a branched 2-surface, or 2-complex. A specific model is given by a partition function that sums over all microscopic degrees of freedom and all (model-dependent) weights on the vertices of the 2-complex. It also sums over all 2-complexes that interpolate between the given in and out 3-geometry states, making spin foams a path-integral approach to quantum gravity. (Markopoulou 2002: 3)

As befits a quantum geometry, the area and volume operators of these models have discrete eigenvalues. It appears that in these 3-geometry states, quantum excitations are one-dimensional spin network edges, where each edge intersecting a surface gives it a quantum of area. To generate enough area for a sheet of paper, it has been calculated that it would require 10^{68} incident edges.

These spin foams are seen as embedded in a differential manifold, but true to Penrose's original vision, some people are approaching spin foams and networks abstractly, that is, without any prior manifold. They encode the representation theory of a group as a quantum field theory, where a duality between geometry and algebra, similar to the knot–quantum group relation, comes into play which, as Baez puts it, 'is exactly the sort of thing one would hope for in a theory of quantum gravity' (1999: 36).

Part of the process of making spin foams into combinatorial TQFTs is to assign labels to faces and maps to edges, combining labels of the faces meeting at the edges. Then a weighted sum is carried out over labellings. But to be 'topological' there naturally have to be severe constraints imposed. As the co-creator of a 4-dimensional TQFT puts it:

> In order to obtain a topologically invariant theory, we need the combination factors to satisfy some equations. The equations they need to satisfy are very algebraic in nature; as we go through different classes of theories in different low dimensions we first rediscover most of the interesting classes of associative algebras, then of tensor categories. (Crane 1994: 125)

In this 'marriage between algebra and topology which underlies TQFT' (Crane 1994: 126), we meet again the same idea of carving out 'natural' algebras via categorification. Categorification corresponds to shifting up a dimension:

> We have discovered that in many situations if a given type of algebraic structure can be used to generate a TQFT in $d = n$ then a categorification can naturally be used to generate a TQFT in $D = n + 1$. (Crane 1995: 9)

For example, in two dimensions semisimple algebras are required. In dimension 3, you must either add comultiplication resulting in Hopf algebras, or categorify to monoidal representation categories. At the next dimension, you complete the square by devising Hopf categories.

I have talked about the three ways of producing a TQFT, but from what we have seen it should be clear that all have a category theoretic flavour. About the combinatorial way, Baez claims:

> In fact, the recipe for amplitudes and the verification of these facts make heavy use of category theory. The same is true for all other theories for which Atiyah's axioms have been verified. For some strange reason, it seems that category theory is precisely suited to explaining what makes a TQFT tick. (Baez 1999: 17)

It would also seem that category theory is likely to benefit from any progress in quantum gravity, because it is heavily implicated in the mathematics equipping the other main contender, namely, string theory and, for instance, its use of derived categories. But, for those who are wary of Planck-scale physics, we do not have to travel to such remote regions to find signs of quantum algebra at work. I stated in chapter 9 that quantum groups would make an excellent topic for a history exemplifying late twentieth-century mathematical physics. Quantum groups emerged as ways to measure symmetry in the physics of certain kinds of two-dimension lattice models and $1 + 1$-dimensional scattering models. For instance, in three-particle interactions, models which are insensitive to the order of the pairwise interactions are organised by the so-called Yang–Baxter equation. In other words, this is a piece of algebra which reflects an equivalence between the following diagrams:

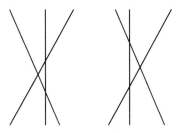

But this diagram corresponds to the third of the Reidermeister moves, the basic moves which allow you to pass between different knot projections of the same knot. This nexus of concepts relating knot theory, $1 + 1$-dimensional scattering models, integrable two-dimensional lattice models of statistical mechanics, and topological quantum field theory in 3 dimensions is currently receiving the categorification treatment, where up a dimension what corresponds to the Yang–Baxter equation is the Zamolodchikov equation.

10.6 CONCLUSION

One can say with little fear of contradiction that in today's philosophy of mathematics, it is the philosophy which dictates the agenda. The issue at stake in this chapter is at what point does it become incumbent upon philosophers to take the reverse attitude and let the mathematics have some say in what is asked of it? Personally I have a very low threshold. I cannot bear to see all this exciting, yet incompletely achieved, mathematics go unnoticed. Some people's threshold is evidently higher, hence the thought to bombard the reader through this chapter with views of this material from enough angles to form the beginnings of at least a dozen doctoral theses.

Consider the position in which Bertrand Russell claimed to have found himself at the start of the twentieth century. He had given up on the idea that British Hegelianism could supply a way of understanding the problems mathematicians face, and he had discerned from the writings of Dedekind,

Weierstrass, Frege and others that a new way of thinking about mathematics was in the process of emerging. This new way of thinking required a new language, and so he rushed to join in the construction of this logic, hopeful that it might revitalise philosophy as a whole. From our vantage point today we can emulate Russell by:

(1) Believing that our current philosophy is not adequate to make proper sense of contemporary mathematics

(2) Trusting that some mathematicians can give us insight into a better philosophical treatment

(3) Believing that the emerging picture will revitalise philosophy.

There are two ways to respond to belief (1). The first way involves taking the philosophy of mathematics in a 'science studies' direction. The material presented in this chapter suggests a range of ways of doing so. A change in the language of mathematics should be expected to reveal previously tacit understandings about mathematics – for example, presuppositions about the use of notation. We also saw how printing technology may have played a significant role in the shape taken by contemporary mathematics. Further, we touched on the changing relationship between physics and mathematics over the past hundred years. Relating to (3), we would hope to contribute to science studies by discovering features which are more prominently displayed by mathematics as a knowledge-producing discipline, e.g., the relationship between content and form, or aesthetics and rationality. According to our attitude to (2), we can be more or less appreciative of what goes on in mathematics departments.

The second way to update belief (1) buys into updated versions of (2) and (3). It is to see a new 'laws of form' or 'laws of thought' emerging which could reinvigorate philosophy by making it think about the nature of a piece of mathematical syntax, the space in which it is written, and what it is supposed to refer to. The philosophy of geometry could be revitalised by the discrete geometry emerging from this programme. While concerning logic, we can already see signs of philosophical activity. Makkai (1999), for example, explains his programme to develop a foundational language out of higher-dimensional algebra which stops you asking 'silly' questions, such as whether two groups are not only isomorphic but also the same. Such a language might be useful for ontologists working outside of mathematics.

Both these ways are surely worth pursuing. Mathematics has been and remains a superb resource for philosophers. Let's not waste it.

Appendix

MONOID

A monoid is a set M with a binary operation, such that
(i) The operation is associative, i.e., $g\cdot(h\cdot k) = (g\cdot h)\cdot k$ for any three elements of M.
(ii) There is an identity element, e, with $e\cdot g = g = g\cdot e$ for all g in M.

GROUP

A group is a monoid $(G, \cdot)$ satisfying the additional condition:
(iii) For each g in G, there is g^{-1} with $g\cdot g^{-1} = e = g^{-1}\cdot g$.
A group is said to *commutative* or *Abelian* if $g\cdot h = h\cdot g$, for all pairs of elements of G.

RING

A ring $R = (R, +, \cdot, 1)$ is a set R with two binary operations, addition and multiplication, and an element 1, such that:
(i) $(R, +)$ is a commutative group.
(ii) $(R, \cdot, 1)$ is a monoid.
(iii) Multiplication is distributive over addition, i.e., $a\cdot(b + c) = a\cdot b + a\cdot c$ and $(a + b)\cdot c = a\cdot c + b\cdot c$.

IDEAL

An ideal I is a subset of a ring R such that if a and b are members of I so is $a + b$, and ra is in I for all r in R. The *principal* ideal generated by x, Ix, is the set of rx with r in R.

FIELD

A field is a non-trivial commutative ring in which every non-zero element has a multiplicative inverse.

ALGEBRA

An algebra over a field k is a vector space A together with two linear maps, an associative multiplication $m: A \otimes A \to A$, and a unit map $\eta: k \to A$, such that $m \cdot (\eta \otimes id_A)$ and $m \cdot (id_A \otimes \eta)$ are scalar multiplication.

COALGEBRA

A coalgebra over a field k is a vector space C together with two linear maps, a comultiplication $\Delta: C \to C \otimes C$ and a counit $\varepsilon: C \to k$, such that:

(i) $(\Delta \otimes id_C) \cdot \Delta = (id_C \otimes \Delta) \cdot \Delta$ (coassociativity);

(ii) $(\varepsilon \otimes id_C) \cdot \Delta = 1_k \otimes id_C$ and $(id_C \otimes \varepsilon) \cdot \Delta = id_C \otimes 1_k$.

BIALGEBRA

$(B, m, \beta, \eta, \Delta, \varepsilon)$ is a bialgebra if (B, m, η) is an algebra, (B, Δ, ε) is a coalgebra, and the algebra and coalgebra structures are compatible, in the sense that:

(i) $\Delta \cdot m = (m \otimes m)(id_H \otimes t \otimes id_H)(\Delta \otimes \Delta)$ where t interchanges the components of the tensor product;

(ii) $\varepsilon \cdot m = \varepsilon \otimes \varepsilon$.

HOPF ALGEBRA

A Hopf algebra is a bialgebra together with an *antipode* $S: H \to H$ such that $m \cdot (S \otimes id_H) \cdot \Delta = \eta \cdot \varepsilon = m \cdot (id_H \otimes S) \cdot \Delta$.

CATEGORY

A category **C** consists of:

(i) A class of objects, $A, B, C, \ldots$

(ii) A class of morphisms (or arrows), $f, g, h, \ldots$, each of which has a domain and codomain which are objects of C. If the domain and codomain of f are A and B, respectively, we may write $f: A \to B$.

(iii) A composition law which takes a pair of *compatible* morphisms (f, g), i.e., with the codomain of f equal to the domain of g, to another morphism: if $f: A \to B$ and $g: B \to C$, then $g \circ f: A \to C$.

These satisfy the following axioms:

(1) Composition is associative.

(2) For each object A there is an identity morphism, 1_A, such that if $f: A \to B$, then $f \circ 1_A = f = 1_B \circ f$ as arrows from A to B.

FUNCTOR

A functor $T: \mathbf{C} \to \mathbf{D}$ is a morphism of categories. In other words, it consists of functions taking objects and morphisms of $\mathbf{C}$, respectively, to the objects and morphisms of $\mathbf{D}$, such that
  (i) If $f: A \to B$, then $T(f): T(A) \to T(B)$.
 (ii) $T(g \circ f) = T(g) \circ T(f)$, whenever $g \circ f$ is defined.
(iii) For all A in C, $T(1_A) = 1_{T(A)}$.

ADJOINT FUNCTORS

Given functors $F: \mathbf{C} \to \mathbf{D}$ and $G: \mathbf{D} \to \mathbf{C}$, we say F is left adjoint to G if there is a bijection, natural in the variables A and B, between morphisms $f: A \to G(B)$ in $\mathbf{C}$ and morphisms $f: F(A) \to B$ in $\mathbf{D}$.

MORITA EQUIVALENCE

Two rings are said to be Morita equivalent if their categories of left modules are equivalent. This is a useful notion which views as 'the same' a ring R and a matrix ring over it, $M_n(R)$. Since even for commutative R, $M_n(R)$ will in general not be commutative, within the category of rings and homomorphisms Morita equivalence cannot be detected, even if we extend it to a strict 2-category with intertwiners acting between arrows. But we can define a bicategory of rings where this equivalence is naturally expressed. The objects in this category are the rings and the arrows between two rings, R and S, are bimodules. The identity maps are formed by canonical bimodules. Horizontal composition corresponds to forming the tensor product. Vertical arrows are R-S linear maps. Then R and S are Morita equivalent if isomorphic in this bicategory. This requires there to be bimodules M and M^{-1} such that

$$S \to M^{-1} \otimes_R M \leftarrow S \cong S \to S \leftarrow S;$$
$$R \to M \otimes_S M^{-1} \leftarrow R \cong R \to R \leftarrow R;$$

each isomorphism provided by inverse vertical arrows. Similar constructions work for C^*-algebras, von Neumann algebras, Lie groupoids, symplectic groupoids and integrable Poisson manifolds. In each case isomorphic objects in the respective bicategories are Morita equivalent.

Bibliography

Abramsky, S., 1997, 'Games in the Semantics of Programming Languages', in *Proceedings of the 11th Amsterdam Colloquium*, P. Dekker, M. Stokhof and Y. Venema (eds.), University of Amsterdam: 1–6

Adams, F., 1978, *Infinite Loop Spaces*, Princeton University Press

Albert, M., 2001, 'Bayesian Learning and Expectations Formation: Anything Goes', in Corfield and Williamson (eds.): 341–62

Alexanderson, G., 2000, *The Random Walks of George Pólya*, The Mathematical Association of America

Alon, A., Bourgain, J., Connes, A., Gromov, M. and Milman, V. (eds.), 2000, *GAFA 2000 Visions in Mathematics: Towards 2000*, Birkhäuser

Arnold, V., 1963, 'Small Denominators and Problems of Stability of Motion', *Russian Mathematical Surveys*, 18(6): 85–192

(ed.), 1988, *Dynamical Systems III*, Springer-Verlag

2000, 'Polymathematics: Is Mathematics a Single Science or a Set of Arts?', in Arnold *et al.* (eds.): 403–16

Arnold, V., Atiyah, M., Lax, P. and Mazur, B. (eds.), 2000, *Mathematics: Frontiers and Perspectives*, American Mathematical Society

Ashtekar, A., 2002, 'Quantum Geometry in Action: Big Bang and Black Holes', preprint math-ph/0202008 available at xxx.arXiv.cornell.edu

Aspray, W. and Kitcher, P. (eds.), 1988, *History and Philosophy of Modern Mathematics*, University of Minnesota Press

Atiyah, M., 1974, 'How Research is Carried Out', *Bulletin of the Institute of Mathematics and its Applications*, 10: 232–4, reprinted in Atiyah (1987): 213–15

1976, 'Global Geometry', *Proceedings of the Royal Society of London A*, 347: 291–9, reprinted in Atiyah (1987): 217–28

1977, 'Trends in Pure Mathematics', in H. Athen and H. Kunle (eds.), *Proceedings of Third ICME*: 61–74, reprinted in Atiyah (1987): 263–76

1978, 'The Unity of Mathematics', *Bulletin of the London Mathematical Society*, 10: 69–76, reprinted in Atiyah (1987): 277–86

1984, 'An Interview with Michael Atiyah', *Mathematical Intelligencer*, 6(1), reprinted in Atiyah (1987): 297–307

1987, *Collected Works, Vol. 1: Early Papers, General Papers*, Oxford University Press

1995, 'Quantum Theory and Geometry', *Journal of Mathematical Physics*, 36(11), 6069–72

Audi, R., 1995, *Cambridge Dictionary of Philosophy*, Cambridge University Press

Bacon, F., 1620, *Novum Organum*, trans. R. Ellis and J. Spedding (eds.), *The Philosophical Works of Francis Bacon*, Routledge, 1905: 212–387

Baez, J., 1999, 'An Introduction to Spin Foam Models of *BF* Theory and Quantum Gravity', preprint gr-qc/9905087 available at xxx.arXiv.cornell.edu

2001a, 'Higher-Dimensional Algebra and Planck-Scale Physics', in Callender and Huggett (eds.): 177–95

2001b, 'Week 169', available at http://math.ucr.edu/home/baez

2002, 'The Octonions', *Bulletin of the American Mathematical Society*, 39(2): 145–205

Baez, J. and Dolan, J., 1995, 'Higher-Dimensional Algebra and Topological Quantum Field Theory', *Journal of Mathematical Physics*, 36: 6073–6105

1999, 'Categorification', in *Higher Category Theory*, E. Getzler and M. Kapranov (eds.), American Mathematical Society: 1–36

Bailey, D. and Borwein, J., 2001, 'Experimental Mathematics: Recent Developments and Future Outlook', in Engquist and Schmid (eds.): 51–66

Bailey, D., Borwein, P. and Plouffe, S., 1997, 'On the Rapid Computation of Various Polylogarithmic Constants', *Mathematics of Computation*, 66(218): 903–13

Bailey, D. and Crandall, R., 2001, 'On the Random Character of Fundamental Constant Expansions', *Experimental Mathematics*, 10(2): 175–90

Barnes, B., Bloor, D. and Henry, J., 1996, *Scientific Knowledge: A Sociological Analysis*, University of Chicago Press

Barnes, E., 1999, 'The Quantitative Problem of Old Evidence', *British Journal for the Philosophy of Science*, 50(2): 249–64

Barrow-Green, J., 1997, *Poincaré and the Three Body Problem*, American Mathematical Society

Bell, J. and Machover, M., 1977, *A Course in Mathematical Logic*, North-Holland

Bellissard, J., 1992, 'Gap Labelling Theorems for Schrödinger Operators', in Waldschmidt *et al.* (eds.): 538–631

van Bendegem, J., 2000 'Analogy and Metaphor as Essential Tools for the Working Mathematician', in F. Hallyn (ed.), *Metaphor and Analogy in the Sciences*, Kluwer: 105–23

Berger, M., 2000, 'Encounter with a Geometer, Part I', *Notices of the American Mathematical Society*, 47(2): 183–94

Bergeron, F., Labelle, G. and Leroux, P., 1997, *Combinational Species and Tree-Line Structures*, Cambridge University Press

Bloor, D., 1976, *Knowledge and Social Imagery*, Routledge

Booss, B. and Bleecker, D. D., 1985, *Topology and Analysis: The Atiyah–Singer Index Theorem and Gauge Theoretic Physics*, Springer-Verlag

Borel, A., 1983, 'Mathematics, Art and Science', *Mathematical Intelligencer*, 5(4): 9–17

Boumans, M., 1999, 'Built-In Justification', in Morgan and Morrison (eds.): 66–96

Brady, G. and Trimble, T., unpublished, 'A String Diagram Calculus for Predicate Logic'

Brandt, H., 1926, 'Über eine Verallgemeinerung des Gruppenbegriffes', *Mathematische Annalen*, 96: 360–6

Breger, H., 1992, 'A Restoration that Failed: Paul Finsler's Theory of Sets', in Gillies (ed.): 248–64

Browder, F. (ed.), 1983, *The Mathematical Heritage of Henri Poincaré*, American Mathematical Society

Brown, J., 1999, *Philosophy of Mathematics: An Introduction to the World of Proofs and Pictures*, Routledge

Brown, R., 1987, 'From Groups to Groupoids: A Brief Survey', *Bulletin of the London Mathematical Society*, 19: 113–34

    1992, 'Out of Line', *Proceedings of the Royal Institution*, 64: 207–43

    1994, 'Higher Order Symmetry of Graphs', *Bulletin of the Irish Mathematical Society*, 32: 46–59

    1999, 'Groupoids and Crossed Objects in Algebraic Topology', *Homology, Homotopy and Applications*, 1(1), 1–78

Bundy, A., 1999, 'Proof Planning Methods as Schemas', *Journal of Symbolic Computation*, 11: 1–25

Butterfield, J. and Isham, C., 2001, 'Spacetime and the Philosophical Challenge of Quantum Gravity', in Callender and Huggett (eds.): 33–89

Callender C. and Huggett N. (eds.), 2001 *Physics meets Philosophy at the Planck Scale*, Cambridge University Press

Carnap, R., 1950, *Logical Foundations of Probability*, 2nd edn., University of Chicago Press, 1963

Carter J., Kauffman, L. and Saito, M., 1996, 'Structures and Diagrammatics of Four Dimensional Topological Lattice Field Theories', preprint math.GT/9806023 available at xxx.arXiv.cornell.edu

Carter, J. and Saito, M., 1998, *Knotted Surfaces and Their Diagrams*, American Mathematical Society

Cartier, P., 1992, 'An Introduction to Zeta Functions', in Waldschmidt *et al.* (eds.): 1–63

    2001, 'A Mad Day's Work: From Grothendieck to Connes and Kontsevich – The Evolution of Concepts of Space and Symmetry', *Bulletin of the American Mathematical Society*, 38(4): 389–408

Cartwright, N., 1999a, *The Dappled World*, Cambridge University Press

    1999b, 'Models and the Limits of Theory: Quantum Hamiltonians and the BCS Model of Superconductivity', in Morgan and Morrison (eds.): 241–81

Chalmers, A., 1982, *What Is This Thing Called Science?*, 2nd edn., Open University Press

Collingwood, R. G., 1940, *An Essay on Metaphysics*, Clarendon Press

    1946, *The Idea of History*, Oxford University Press

    1999, *The Principles of History*, Clarendon Press

Collins, H. and Kusch, M., 1998, *The Shape of Actions*, MIT Press

Colton, S., 1999, 'Refactorable Numbers – A Machine Invention', *Journal of Integer Sequences*, 2, article 99.1.2

Connes, A., 1994, *Noncommutative Geometry*, Academic Press

Connes, A., Lichnerowicz, A. and Schützenberger, M., 2000, *Triangle of Thoughts*, trans. J. Gage, American Mathematical Society

Corfield, D. and Williamson, J. (eds.), 2001, *Foundations of Bayesianism*, Kluwer

Corry, L., 1996, *Modern Algebra and the Rise of Mathematical Structures*, Birkhäuser

da Costa, N. and French, S., 2000, 'Theories, Models and Structures: Thirty Years On', *Philosophy of Science*, 67, Supplement: Proceedings of PSA 1998: S116–S127

Crane, L., 1994, 'Topological Field Theory as the Key to Quantum Gravity', in J. Baez (ed.), *Knots and Quantum Gravity*, Oxford University Press: 121–32

 1995, 'Clock and Category: Is Quantum Gravity Algebraic?', *Journal of Mathematical Physics*, 36: 6180–93

Crowe, M., 1975, 'Ten "Laws" Concerning Patterns of Change in the History of Mathematics', *Historia Mathematica*, 2: 161–6, reprinted in Gillies (ed.) (1992): 15–20

 1990, 'Duhem and the History and Philosophy of Mathematics', *Synthese*, 83: 431–47

Crowell, H. and Fox, R., 1977, *Introduction to Knot Theory*, Springer-Verlag

Davis, P. J. and Hersh, R., 1981, *The Mathematical Experience*, Birkhäuser

Deninger, C., 1994, 'Evidence for a Cohomological Approach to Analytic Number Theory', in A. Joseph *et al.* (eds.), *First European Congress of Mathematics, Vol. 1*, Birkhäuser: 491–510

Dieudonné, J., 1969, 'Richard Dedekind', *Encyclopaedia Universalis, Vol. 5*: pp. 373–5

 1972, 'The Historical Development of Algebraic Geometry', *American Mathematical Monthly*, 79: 827–66

 1975a, 'Poincaré', in C. C. Gillispie (ed.), *Dictionary of Scientific Biography*, Charles Scribner's Sons: 52

 1975b, 'L'Abstraction et l'intuition mathématique', *Dialectica*, 29(1): 39–54

 1989, *A History of Algebraic and Differential Topology: 1900–1960*, Birkhäuser

Earman, J., 1992, *Bayes or Bust?: A Critical Examination of Bayesian Confirmation Theory*, MIT Press

Edwards, H. M., 1977, *Fermat's Last Theorem – A Genetic Introduction to Algebraic Number Theory*, Springer

 1989, *Divisor Theory*, Birkhäuser

Eilenberg, S. and Steenrod, N., 1952, *Foundations of Algebraic Topology*, Princeton University Press

Eisenbud, D., 2001, 'Mathematics Comes from Many Sources', in Engquist and Schmid (eds.): 647–54

Engquist, B. and Schmid, W. (eds.), 2001, *Mathematics Unlimited – 2001 and Beyond*, Springer-Verlag

Ewald, W. (ed.), 2000, *From Kant to Hilbert: A Sourcebook in the Foundations of Mathematics*, Oxford University Press

Fallis, D., 1997, 'The Epistemic Status of Probabilistic Proof', *Journal of Philosophy*, 94(4): 165–86

Feferman, S., 1981, 'The Logic of Mathematical Discovery vs. the Logical Structure of Mathematics', in P. D. Asquith and I. Hacking (eds.), *PSA 1978*, 2, Philosophy of Science Association: 309–27

Ferreirós, J., 2001 'The Road to Modern Logic – An Interpretation', *Bulletin of Symbolic Logic*, 7(4): 441–84

Feyerabend, P., 1976, 'On the Critique of Scientific Reason', in *Problems of Empiricism*, Philosophical Papers, 2, Cambridge University Press, ch. 10

Field, H., 1980, *Science Without Numbers*, Blackwell

de Finetti, B., 1974, *Theory of Probability: A Critical Introductory Treatment*, 1, trans. by A. Machi and A. Smith, Wiley

Fitelson, B., 1998, 'Using *Mathematica* to Understand the Computer Proof of the Robbins Conjecture', *Mathematica in Education and Research*, 7(1): 17–26

Franklin, A., 1986, *The Neglect of Experiment*, Cambridge University Press

Freed, D. and Uhlenbeck, K. (eds.), 1995, *Geometry and Quantum Field Theory*, American Mathematical Society

Friedman, M., 1992, *Kant and the Exact Sciences*, Harvard University Press
2001, *Dynamics of Reason*, CSLI Publications

Galison, P., 1987, *How Experiments End*, University of Chicago Press
1997, *Image and Logic: A Material Culture of Microphysics*, University of Chicago Press

Gannon, T., 2001, 'Postcards from the Edge, or Snapshots of the Theory of Generalised Moonshine', preprint QA/0109067 available at xxx.arXiv.cornell.edu

Gardner, H., 1983, *Frames of Mind*, Paladin

Gilbert, G. and Mulkay, M., 1984, *Opening Pandora's Box: A Sociological Analysis of Scientists' Discourse*, Cambridge University Press

Gilbert, N. and Porter, T., 1994, *Knots and Surfaces*, Oxford University Press

Gillies, D. (ed.), 1992, *Revolutions in Mathematics*, Oxford University Press
1996, *Artificial Intelligence and Scientific Method*, Oxford University Press

Giorello, G., 1980, 'Intuition and Rigor: Some Problems of a "Logic of Discovery" in Mathematics', in M. L. Dalla Chiara (ed.), *Italian Studies in the Philosophy of Science*, Reidel, 113–35

Giventhal, A., 1996, 'Equivariant Gromov–Witten Invariants', *International Mathematical Research Notices*, 13: 613–63

Glas, E., 1989, 'Testing the Philosophy of Mathematics in the History of Mathematics, Part II: The Similarity Between Mathematical and Scientific Growth of Knowledge', *Studies in History and Philosophy of Science*, 20: 157–74
1993, 'Mathematical Progress: Between Reason and Society, I. The Methodological Model and its Alternatives', *Journal of General Philosophy of Science*, 1: 43–62
1995, 'Kuhn, Lakatos, and the Image of Mathematics', *Philosophia Mathematica*, 3: 225–47

Gowers, W. T., 2000a, 'The Two Cultures of Mathematics', in Arnold *et al.* (eds.): 65–78

2000b, 'Rough Structure and Classification', in Alon *et al*. (eds.): 79–117

Grosholz, E., 1985, 'Two Episodes in the Unification of Logic and Topology', *British Journal for the Philosophy of Science*, 36: 147–57

Hacking, I., 1967, 'Slightly More Realistic Personal Probability', *Philosophy of Science*, 34: 311–25

1979, 'Lakatos's Philosophy of Science', *British Journal for the Philosophy of Science*, 30: 381–410

1983, *Representing and Intervening: Introductory Topics in the Philosophy of Natural Science*, Cambridge University Press

1992, '"Style" for Historians and Philosophers', *Studies in History and Philosophy of Science*, 23: 1–20

1996, 'The Disunities of the Sciences', in P. Galison and D. Stump (eds.), *The Disunity of Science: Boundaries, Contexts and Power*, Stanford University Press

1999, *The Social Construction of What?*, Harvard University Press

Hallett, M., 1979, 'Toward a Theory of Mathematical Research Programmes', *British Journal for the Philosophy of Science*, 30: 1–25, 135–59

Hardy, G. H., 1940, *A Mathematician's Apology*, Cambridge University Press

Hasse, H., 1980, *Number Theory*, Springer–Verlag

Hempel, C., 1945, 'On the Nature of Mathematical Truth', in P. Benacerraf and H. Putnam (eds.), *Philosophy of Mathematics: Selected Readings*, Prentice-Hall, 1964

Herrnstein Smith, B. and Plotnitsky, A. (eds.), 1997, *Mathematics, Science, and Postclassical Theory*, Duke University Press

Hersh, R., 1979, 'Some Proposals for Reviving the Philosophy of Mathematics', *Advances in Mathematics*, 31(1): 31–50

1991, 'Mathematics has a Front and a Back', *Synthèse*, 88: 127–33

Hesse, M., 1974, *The Structure of Scientific Inference*, Macmillan

Hestenes, D., 1986, 'A Unified Language for Mathematics and Physics', in J. Chisholm and A. Commons (eds.), *Clifford Algebras and their Applications in Mathematical Physics*, Reidel: 1–23

Higgins, P., 1971, *Notes on Categories and Groupoids*, Van Nostrand Reinhold Co.

Hirsch, M., 1994, 'Responses to "Theoretical Mathematics", by A. Jaffe and F. Quinn', *Bulletin of the American Mathematical Society*, 30(2): 187–91

Hitchin, N., 2001, 'Global Differential Geometry', in Engquist and Schmid (eds.): 577–91

Howson, C., 2000, *Hume's Problem*, Oxford University Press

Howson, C. and Urbach, P., 1989, *Scientific Reasoning: The Bayesian Approach*, Open Court

Hughes, R., 1999, 'The Ising Model, Computer Simulation, and Universal Physics', in Morgan and Morrison (eds.): 97–145

Hume, D., 1739, *A Treatise of Human Nature*, Clarendon Press, 1978

Jackendoff, R., Bloom, P. and Wynn, K. (eds.), 1999, *Language, Logic, and Concept: Essays in Memory of John Macnamara*, MIT Press

Jackson, A., 2001, 'Interview with Raoul Bott', *Notices of the American Mathematical Society*, 48(4): 374–82

Jaffe, A. and Quinn, F., 1993, '"Theoretical Mathematics": Towards a Cultural Synthesis of Mathematics and Theoretical Physics', *Bulletin of the American Mathematical Society*, 29: 1–13

Jardine, N., 2000, *The Scenes of Inquiry*, revised edn., Oxford University Press

Jaynes, E., forthcoming, *Probability Theory: The Logic of Science*, available at http://bayes.wustl.edu

Jeffrey, R., 1992, *Probability and the Art of Judgment*, Cambridge University Press

Johnstone, P., 1977, *Topos Theory*, Academic Press

   1982, *Stone Spaces*, Cambridge University Press

   (1984), 'Open Locales and Exponentiation', in J. W. Gray (ed.), *Mathematical Applications of Category Theory*, American Mathematical Society

Joyal, A. and Street, R., 1991, 'The Geometry of Tensor Calculus I', *Advances in Mathematics*, 88(1): 55–112

Kac, M., Rota, G.-C. and Schwartz, J., 1986, *Discrete Thoughts: Essays in Mathematics, Science, and Philosophy*, Birkhäuser

Kaminker, J., 1998, 'Conference Announcement: Groupoids in Physics, Analysis and Geometry', at http://www.ams.org/meetings/src-kaminker.html (8 December 1998)

Kapranov, M., 1995, 'Analogies between the Langlands Correspondence and Topological Quantum Field Theory', in S. Gindikin, J. Lepowsky, and R. Wilson, (eds.), *Functional Analysis on the eve of the 21st Century*, 1, Birkhäuser: 119–51

   2001, 'Review of Gröbner Deformations of Hypergeometric Differential Equations', *Bulletin of the American Mathematical Society*, 38(4): 481–8

Katz, N. and Sarnak, P., 1999, 'Zeroes of Zeta Functions and Symmetry', *Bulletin of the American Mathematical Society*, 36(1): 1–26

Kauffman, L., 1991, *Knots and Physics*, World Scientific

   2000, 'The Robbins Problem – Computer Proofs and Human Proofs', available at http://www.math.uic.edu/~kauffman/Papers.html

Kerber, M., Kohlhase, M. and Sorge, V., 1998, 'Integrating Computer Algebra into Proof Planning', *Journal of Automated Reasoning*, 21(3): 327–55

Kitcher, P., 1983, *The Nature of Mathematical Knowledge*, Oxford University Press

   1993, *The Advancement of Science*, Oxford University Press

Klainerman, S., 2000, 'PDE as a Unified Subject', in Alon *et al.* (eds.): 279–315

Kline, M., 1972, *Mathematical Thought from Ancient to Modern Times*, Oxford University Press

Koetsier, T., 1991, *Lakatos' Philosophy of Mathematics: A Historical Approach*, North-Holland

Krattenthaler, C., 1999, 'Advanced Determinant Calculus', *Séminaire Lotharingien de Combinatoire*, 42: article B42q

Kreimer, D., 2000, *Knots and Feynman Diagrams*, Cambridge University Press

Kreisel, G. and MacIntyre, A., 1982, 'Constructive Logic Versus Algebraization I', in A. Troelstra and D. van Dalen (eds.), *The L. E. J. Brouwer Centenary Colloquium*, North-Holland

Kuhn, T., 1962, *The Structure of Scientific Revolutions*, University of Chicago Press

   1977, *The Essential Tension*, University of Chicago Press

Kumpera, A., 1988, Review of [Mackenzie 1987], *Bulletin of the American Mathematical Society*, 19(1): 358–62

Ladyman, J., 1998, 'What is Structural Realism?', *Studies in History and Philosophy of Science*, 29: 409–24

Lagarias, J., 1992, 'Number Theory and Dynamical Systems', in S. Burr (ed.), *The Unreasonable Effectiveness of Number Theory*, American Mathematical Society: 35–72

Lakatos, I., 1968, 'Changes in the Problem of Inductive Logic', in Lakatos (1978b): 128–200

    1976, *Proofs and Refutations, The Logic of Mathematical Discovery*, J. Worrall and E. Zahar (eds.), Cambridge University Press

    1978a, *The Methodology of Scientific Research Programmes, Philosophical Papers, Vol. 1*, J. Worrall and G. Currie (eds.), Cambridge University Press

    1978b, *Mathematics, Science and Epistemology, Philosophical Papers, Vol. 2*, J. Worrall and G. Currie (eds.), Cambridge University Press

Lang, S., 1970, *Algebraic Number Theory*, Addison-Wesley

Larvor, B., 1998, *Lakatos: An Introduction*, Routledge

    2001, 'What is Dialectical Philosophy of Mathematics?' *Philosophia Mathematica*, 9(2): 212–29

Laudan, L., 1984, *Science and Values*, University of California Press

Laugwitz, D., 1999, *Bernhard Riemann, 1826–1866: Turning Points in the Conception of Mathematics*, Birkhäuser

Lawvere, F. W. and Schanuel, S., 1997, *Conceptual Mathematics, A First Introduction to Categories*, Cambridge University Press

Lefschetz, S., 1971, *Selected Papers*, Chelsea

Leinster, T., 2002, 'A Survey of Definitions of n-Category', *Theory and Applications of Categories*, 10: 1–70, preprint math.CT/0107188 available at xxx.arXiv.cornell.edu

Macintyre, A., 1989, 'Trends in Logic', in R. Ferro, C. Bonotto, S. Valentini and A. Zarnardo (eds.), *Logic Colloquium '88*, North-Holland

MacKenzie, D., 2001, *Mechanizing Proof: Computing, Risk, and Trust*, MIT Press

Mackenzie, K., 1987, *Lie Groupoids and Lie Algebroids in Differential Geometry*, Cambridge University Press

    1992, 'Double Lie Algebroids and Second-Order Geometry, I', *Advances in Mathematics*, 92(2): 180–239

    1995, 'Lie Algebroids and Lie Pseudogroups', *Bulletin of the London Mathematical Society*, 27: 97–147

Mackey, G. W., 1992, *The Scope and History of Commutative and Noncommutative Harmonic Analysis*, American Mathematical Society

Mac Lane, S., 1971, *Categories for the Working Mathematician*, Springer-Verlag

    1978, 'Origins of the Cohomology of Groups', *L'Enseignement Mathématique*, XXIV: fasc. 1–2

    1986, *Mathematics: Form and Function*, Springer–Verlag

    1988, 'Concepts and Categories in Perspective', in P. Duren, R. Askey and U. Merzbach (eds.), *A Century of Mathematics in America, Part I*, American Mathematical Society: 323–65

1992, 'The Protean Character of Mathematics', in J. Echeverria, A. Ibarra and T. Mormann (eds.), *The Space of Mathematics: Philosophical, Epistemological and Historical Explorations*, Walter de Gruyter: 3–13

Macnamara, J. and Reyes, G. (eds.), 1994, *The Logical Foundations of Cognition*, Oxford University Press

Maddy, P., 1997, *Naturalism in Mathematics*, Clarendon Press

Majid, S., 1995, *Foundations of Quantum Groups*, Cambridge University Press

Makkai, M., 1999 'On Structuralism in Mathematics', in R. Jackendoff *et al.* (eds.): 43–66

Mancosu, P., 1996, *Philosophy of Mathematics and Mathematical Practice in the Seventeenth Century*, Oxford University Press

Markopoulou F., 2002 'Coarse Graining in Spin Foam Models', preprint gr-qc/0203036 available at xxx.arXiv.cornell.edu

Marquis, J.-P., 1997, 'Abstract Mathematical Tools and Machines for Mathematics', *Philosophia Mathematica*, 5: 250–72

Martin, U., 1999, 'Computers, Reasoning and Mathematical Practice', in *Proceedings of the 1997 NATO ASI Summer School on Logic and Computation*, Springer Verlag

McIntyre, M. and Trimble, T., forthcoming, 'The Geometry of Gray-Categories', *Advances in Mathematics*

McLarty, C., 1990, 'The Uses and Abuses of the History of Topos Theory', *British Journal for the Philosophy of Science*, 41: 351–75

2000, 'Voir-dire in the Case of Mathematical Physics', in E. Grosholz and H. Breger (eds.), *The Growth of Mathematical Knowledge*, Kluwer: 269–80

Mehrtens, H., 1976, 'T. S. Kuhn's Theories and Mathematics', *Historia Mathematica*, 3: 297–320, reprinted as ch. 2 of Gillies (ed.) (1992)

1990, *Moderne-Sprache-Mathematik*, Suhrkamp

1992, *Appendix (1992): Revolutions Reconsidered*, in Gillies (ed.) (1992): 42–8

Melis, E., 1998, 'The Heine–Borel Challenge Problem: In Honor of Woody Bledsoe', *Journal of Automated Reasoning*, 20(3): 255–82

van Mill, J. and Reed, G. (eds.), 1990, *Open Problems in Topology*, North Holland–Elsevier

Mill, J. S., 1843, *A System of Logic, Ratiocinative and Inductive*, 3rd edn., Longmans, Green & Co.

Moerdijk, I. and Reyes, G., 1991, *Models for Smooth Infinitesimal Analysis*, Springer-Verlag

Morgan, M. and Morrison, M. (eds.), 1999, *Models as Mediators*, Cambridge University Press

Morrow, G., 1970, *Proclus: A Commentary on the First Book of Euclid's Elements*, Princeton University Press

Muggleton, S., 1999, 'Scientific Knowledge Discovery using Inductive Logic Programming', *Communications of the ACM*, 42(11): 42–6

Needham, T., 1998, *Visual Complex Analysis*, Oxford University Press

Netz, R., 1999, *The Shaping of Deduction in Greek Mathematics*, Cambridge University Press

Okasha, S., 2000, 'Van Fraassen's Critique of Inference to the Best Explanation', *Studies in History and Philosophy of Science*, 31(4): 691–710

Oldfield, A., 1995, 'Metaphysics and History in Collinwood's Thought', in D. Boucher, J. Connelly and T. Modood (eds.), *Philosophy, History and Civilisation: Interdisciplinary Perspectives on R. G. Collingwood*, Univerity of Wales Press: 182–202

Papert, S., 1978, 'The Mathematical Unconscious', in J. Wechsler (ed.), *On Aesthetics in Science*, MIT Press: 105–20

Paris, J., Watton, P. and Wilmers, G., 2000, 'On the Structure of Probability Functions in the Natural World', *International Journal of Uncertainty, Fuzziness and Knowledge-Based Systems*, 8: 311–29

Pearl, J., 1988, *Probabilistic Reasoning in Intelligent Systems*, Morgan Kaufman
2000, *Causality: Models, Reasoning and Inference*, Cambridge University Press
2001, 'Bayesianism and Causality, or, Why I am Only a Half-Bayesian', in Corfield and Williamson (eds.): 19–36

Penrose, R. and Rindler, W., 1984, *Spinors and Space-Time Volume 1: Two-Spinor Calculus and Relativistic Fields*, Cambridge University Press

Pickering, A., 1997, 'Concepts and the Mangle of Practice: Constructing Quaternions', in Herrnstein Smith and Plotnitsky (eds.): 40–82

Poincaré, H., 1895, *Analysis Situs*, Oeuvres, 6, Gauthier-Villars, 1953: 193–288
1905, *Science and Hypothesis*, Dover
1921, *Analyse de ses travaux scientifique*, Acta Mathematica, 38, reprinted in Browder: 257–357

Polanyi, M., 1958, *Personal Knowledge: Towards a Post-Critical Philosophy*, Routledge & Kegan Paul

Pólya, G., 1941, 'Heuristic Reasoning and the Theory of Probability', *American Mathematical Monthly*, 48(7): 450–65
1954a, *Mathematics and Plausible Reasoning: Induction and Analogy in Mathematics, Vol. 1*, Princeton University Press
1954b, *Mathematics and Plausible Reasoning: Patterns of Plausible Inference, Vol. 2*, Princeton University Press

Popper, K., 1963, *Conjectures and Refutations*, Routledge & Kegan Paul

Propp, J., 1999, 'Enumeration of Matchings: Problems and Progress', in L. Billera (ed.), *New Perspectives in Geometric Combinatorics*, Cambridge University Press: 255–91

Quine, W. V., 1955, *Mathematical Logic*, Harvard University Press
1977, 'Review of Proofs and Refutations', *British Journal for the Philosophy of Science*, 28: 81–2

Rabin, J., 1995, 'Introduction to Quantum Field Theory for Mathematicians', in Freed and Uhlenbeck (eds.): 183–269

Rav, Y., 1999, 'Why Do We Prove Theorems?', *Philosophia Mathematica*, 7: 5–41

Ravenel, D., 1986, *Complex Cobordism and Stable Homotopy Groups of Spheres*, Academic Press

Reid, M., 1988, *Undergraduate Algebraic Geometry*, Cambridge University Press

Restivo, S., 1992, *Mathematics in Society and History: Sociological Inquiries*, Kluwer

Reyes, G., 1980, 'Logic and Category Theory', in E. Agazzi (ed.), *Modern Logic: A Survey*, Reidel

Roberts, D., 1973, *The Existential Graphs of Charles S. Peirce*, Mouton

Rosenkrantz, R., 1977, *Inference, Method, and Decision: Towards a Bayesian Philosophy of Science*, Reidel

Rota, G.-C., 1991, 'The Concept of Mathematical Truth', *Review of Metaphysics*, 44: 483–94

Rotman, B., 1997, 'Thinking Dia-grams: Mathematics, Writing, and Virtual Reality', in Herrnstein Smith and Plotnitsky (eds.): 17–39

Rovelli, C., 2001, 'Quantum Spacetime: What do we Know?' in Callender and Huggett (eds.): 101–22

Ruelle, D., 1988, 'Is our Mathematics Natural? The Case of Equilibrium Statistics', *Bulletin of the American Mathematical Society*, 19: 259–68

1991, *Chance and Chaos*, Princeton University Press

Sharpe, R., 1997, *Differential Geometry: Cartan's Generalization of Klein's Erlangen Program*, Springer

Sklar, L., 2000, 'Interpreting Theories: The Case of Statistical Mechanics', *British Journal for the Philosophy of Science*, 51: 129–42

Solomon, R., 2001, 'The Classification of the Finite Simple Groups', *Bulletin of the American Mathematical Society*, 38(3), 315–52

Spencer-Brown, G., 1969, *Laws of Form*, Allen & Unwin

Stanley, R., 1999, *Enumerative Combinatorics*, Vol. 2, Cambridge University Press

Stark, H., 1992, *Galois Theory, Algebraic Numbers and Zeta Functions*, in Waldschmidt *et al.* (eds.): 313–89

Steiner, M., 1975, *Mathematical Knowledge*, Cornell University Press

1983, 'The Philosophy of Imré Lakatos', *Journal of Philosophy*, 80(9): 502–21

1998, *The Applicability of Mathematics as a Philosophical Problem*, Harvard University Press

Stewart, I., 1987, *The Problems of Mathematics*, Oxford University Press

1997, *Does God Play Dice?: The New Mathematics of Chaos*, 2nd edn., Penguin

Sussman, G. and Wisdom, J., 1992, 'Chaotic Evolution of the Solar System', *Science*, 257: 56–62

Tappenden, J., 1995, 'Extending Knowledge and "Fruitful Concepts": Fregean Themes in the Foundations of Mathematics', *Nous*, 29(4): 427–67

Taylor, P., 1999, *Practical Foundations of Mathematics*, Cambridge University Press

Thomas, R., 1998, 'An Update on the Four-Color Theorem', *Notices of the American Mathematical Society*, 45(7): 848–59

Thurston, W., 1982, 'Three Dimensional Manifolds, Kleinian Groups and Hyperbolic Geometry', *Bulletin of the American Mathematical Society*, 6(3): 357–81

1994, 'On Proof and Progress in Mathematics', *Bulletin of the American Mathematical Society*, 30(2): 161–77

Vafa, C., 1998, 'Geometric Physics', *Documenta Mathematica*, Extra Volume ICM, I: 537–56

2000, 'On the Future of Mathematics/Physics Interaction', in Arnold *et al.* (eds.): 321–8

Waldschmidt, M., Moussa, P., Luck, J.-M. and Itzykson C. (eds.), 1992, *From Number Theory to Physics*, Springer-Verlag

Weil, A., 1940, 'Une lettre et un extrait de lettre à Simone Weil', in *Collected Papers*, 1: 244–55

1950, 'Number-Theory and Algebraic Geometry', in *Collected Papers*, 2: 90–100

1960, 'De la Métaphysique aux Mathématiques', in *Collected Papers*, 2: 408–12

1974, 'Two Lectures on Number Theory, Past and Present', in *Collected Papers*, 3: 279–302

1979, *Collected Papers*, Springer-Verlag

Weinstein, A., 1996, 'Groupoids: Unifying Internal and External Symmetry', *Notices of the American Mathematical Society*, 43(7): 744–52

Weyl, H., 1932, 'Topology and Abstract Algebra as Two Roads of Mathematical Comprehension', trans. A. Shenitzer, in *American Mathematical Monthly*, 102(5): 453–60, and 102(7): 646–51

1940, *Algebraic Number Theory*, Princeton University Press

1951, 'A Half Century of Mathematics', *American Mathematical Monthly*, 58: 523–53

Wilson, M., 1999, 'To Err is Humean', *Philosophia Mathematica*, 7(3): 247–57

Wise, M., 1979, 'William Thomson's Mathematical Route to Energy Conservation: A Case Study of the Role of Mathematics in Concept Formation', *Historical Studies of the Physical Sciences*, 10, JHU Press

Wittgenstein, L., 1978, *Remarks on the Foundations of Mathematics*, revised edn., MIT Press

Wos, L., 1988, *Automated Reasoning: 33 Basic Research Problems*, Prentice-Hall

Wos, L. and Fitelson, B., forthcoming, 'The Automation of Sound Reasoning and Successful Proof Finding', to appear in D. Jacquette (ed.), *Blackwell Companion to Philosophical Logic*

Wos, L. and Pieper, G. W., 1999, *A Fascinating Country in the World of Computing: Your Guide to Automated Reasoning*, World Scientific

Wussing, H., 1984, *Genesis of the Abstract Group*, MIT Press

Zahar, E., 1973, 'Why did Einstein's Programme Supersede Lorenz's?', *British Journal for the Philosophy of Science*, 24: 95–123, 223–62

1983, 'Logic of Discovery or Psychology of Invention?', *British Journal for the Philosophy of Science*, 34: 243–61

1989, *Einstein's Revolution: A Study in Heuristic*, Open Court

Index

Lightning Source UK Ltd.
Milton Keynes UK
16 March 2011